Computer Engineer's
Pocket Book

Computer Engineer's Pocket Book

Michael Tooley

Heinemann : London

Heinemann Professional Publishing Ltd
22 Bedford Square, London WC1B 3HH

LONDON   MELBOURNE   JOHANNESBURG   AUCKLAND

First published 1987
Reprinted 1987

© Copyright Michael H. Tooley 1987

All rights reserved. No part of this publication may be
reproduced, stored in a retrieval system, or transmitted, in any
form or by any means, electronic, mechanical, photocopying,
recording or otherwise, without the permission of the publisher
and the copyright owner.

British Library Cataloguing in Publication Data

Tooley, Michael H.
  Computer engineer's pocket book.
  1. Computer engineering
  I. Title
  621.39   TK7885

ISBN 0 434 91967 5

Typeset by Servis Filmsetting Ltd, Manchester
Printed by Butler & Tanner Ltd, Frome

Preface

With the advent of the information technology revolution we are witnessing the convergence of three disciplines: computing, electronics and telecommunications. For us to be able fully to exploit the potential of microprocessors and microcomputers in what remains of the twentieth century, it will become increasingly necessary to abandon the old, and somewhat rigid, boundaries which have until now existed between 'hardware' and 'software'.

Within industry this trend has already manifested itself in the shape of a growing demand for 'software engineers'. But who, or what, are these crusaders of the information technology movement?

They are certainly not primarily programmers, nor are they exclusively electronic engineers. Their talents lie with the integration of software and hardware into fully functional and fully optimized systems.

The precise skills required are hard to define in just a few words but essentially they centre on an awareness of and familiarity with electronic and microelectronic circuitry, coupled with a detailed knowledge of programming in either assembly language or an appropriate high level language. In addition, some knowledge of computer interfacing and communications is highly desirable.

This book aims to provide the sort of everyday information required by such individuals, but it should also be of value to hardware and software specialists. Indeed, anyone engaged in regular use of a computer or microcomputer system at more than just the applications level should find something of value contained herein.

This book cannot replace standard texts or detailed specifications. It does, however, cover a vast range of subjects at a practical level with, where appropriate, some explanatory text. It has also not been designed with readability in mind; rather the aim has been that of presenting

Preface

information in the most concise manner and in a form which can be readily accessed.

Finally, one brief word of advice to the reader. Don't be content to leave this book on the shelf—it should form part of your everyday 'toolkit'. If you use it in much the same manner as your trusty pocket calculator then it will have achieved its aim!

Michael Tooley

Contents

Abbreviations (general) 9
Abbreviations commonly used in pin connection data, etc. 17
Manufacturers' prefixes for semiconductor devices 20
Integrated circuit technologies 22
Scale of integration 23
Basic logic gates 24
Logic circuit equivalents 25
Positive and negative logic equivalents 26
Mixed logic equivalents 26
Typical CMOS and TTL gate circuits 27
TTL and CMOS device coding 27
74 series 29
4000 series 48
4500 series 51
Electrical characteristics of typical logic gates 54
Fan-in and fan-out of logic gates 54
Standard TTL load 55
TTL input and output current 55
Interconnecting TTL families 55
Logic levels and noise margins for CMOS and TTL 56
Boolean algebra 56
Karnaugh maps 57
Power supplies 59
Interfacing logic families 60
Microcomputer architecture 61
Simplified model of a microcomputer system 62
Typical 16-bit microcomputer system 63
Memory maps 63
Internal architecture of a microprocessor 65
Some typical CPUs 67
CPU data 88
Support devices 89
I/O control methods 91
Simple parallel I/O interface 92
Programmable parallel interface devices 92
Programmable serial interface devices 96
Basic cell configuration of semiconductor memories 99
Semiconductor random access memory 99
Semiconductor read only memory 103
Storage capacities of mass memories 107
Magnetic disk storage 107
Magnetic recording techniques 112
IBM 3740 disk format 112
Disk drive mechanics 115
Floppy disk controllers 116
Powers of two 122
Decimal/hexadecimal/octal/binary/ASCII conversion table 122
ASCII control characters 127
Divisors of 255 with remainders 128
Divisors of 256 with remainders 129
Flowchart symbols 130
Operating systems 130
Software tools 134
Languages 138
Video display processing 145

Video resolution 147
Typical video display ASCII character set 148
Typical video standards 149
Typical video waveforms 150
SCART connector pin connections 152
S-100 bus 152
Intel Multibus 156
IBM PC expansion bus 158
Centronics printer interface 160
IEEE-488/GPIB bus 162
Serial data transmission 165
RS-232C/CCITT V24 166
S5/8 interface 172
KERMIT 173
Useful interface circuits 175
Resistor colour code 179
Capacitor colour code 181
Cassette drive stock faults 182
Disk drive stock faults 184
Printer stock faults 186
Monitor stock faults 190
Typical adjustment procedure for a monochrome monitor 194
Index 199

Abbreviations (general)

a.c.	Alternating current
A	Ampere
A/D	Analogue to digital
ACC	Accumulator
ACIA	Asynchronous communications interface adaptor
ACK	Acknowledge
ACU	Automatic calling unit
ADC	Analogue to digital converter
ADCCP	Advanced data communication control procedure
AES	Application environment services
AFIPS	American Federation of Information Processing Societies
ALU	Arithmetic logic unit
AM	Amplitude modulation
ANSI	American National Standards Institution
APPS	Automatic parts programming system
APT	Automatically programmed tools
APU	Arithmetic processor unit
AQL	Acceptable quality level
ARQ	Automatic repeat request
ARU	Audio response unit
ASCII	American Standard Code for Information Interchange
ASM	Assembler
ASR	Automatic send/receive
ATE	Automatic test equipment
ATG	Automatic test generation
AUX	Auxiliary
B	Battery
BABT	British Approvals Board for Telecommunications
BBD	Bucket brigade device
BBS	Bulletin board service
BBT	Bit block transfer
BCCD	Buried channel charge-coupled device
BCD	Binary coded decimal
BCS	British Computer Society
BDOS	Basic disk operating system
BIOS	Basic input/output system
BIU	Bus interface unit
BOS	Business operating system
BPS	Bits per second
BS	Backing store
BSC	Binary synchronous communication
BSI	British Standards Institution
BTAM	Basic telecommunications access method
BV	Bus vectored
c	Centi ($\times 10^{-2}$)
C	Capacitor
CAD	Computer aided design
CAI	Computer aided instruction
CAL	Computer aided learning
CAM	Computer aided manufacture
CCD	Charge-coupled device
CCITT	International Telegraph and Telephone Consultative Committee
CCP	Console command processor

10 Abbreviations (general)

CCR	Condition code register
CERDIP	Ceramic dual-in-line package
CLK	Clock
CML	Current mode logic
CMOS	Complementary metal oxide semiconductor
CSMA	Carrier sense multiple access
CNC	Computer numerical control
CODASYL	Conference on Data Systems Languages
CODEC	Coder/decoder
COM	Computer output to microfilm
CP/M	Control Program for Microcomputers
CPE	Central processing element
CPM	Cards per minute
CPS	Characters per second
CPU	Central processing unit
CR	Card reader
CR	Carriage return
CRC	Cyclic redundancy check
CROM	Control read only memory
CRT	Cathode ray tube
CRTC	Cathode ray tube controller
CU	Control unit
CWP	Communicating word processor
d.c.	Direct current
D	Diode
D/A	Digital to analogue
DAC	Digital to analogue converter
DAC	Data acquisition and control
DAL	Data access line
DAR	Data access register
DART	Dual asynchronous receiver/transmitter
DASM	Direct access storage medium
DASS	Digital access signalling system
DBMS	Database management system
DC	Don't care
DCE	Data communication equipment
DCS	Data carrier system
DCTE	Data circuit terminating equipment
DCTL	Direct coupled transistor logic
DDL	Data description language
DEMUX	Demultiplexer
DIB	Data input bus
DIL	Dual-in-line
DIN	German Standards Institute
DIP	Dual-in-line package
DL	Diode logic
DMA	Direct memory access
DMAC	Direct memory access controller
DMM	Digital multimeter
DMOS	Double diffused metal oxide semiconductor
DMS	Data management system
DNC	Direct numerical control
DOB	Data output bus
DOS	Disk operating system
DP	Data processing
DPM	Digital panel meter
DPMA	Data Processing Management Association
DPNSS	Digital private network signalling system
DPU	Display processing unit

DRAM	Dynamic random access memory
DSDD	Double-sided double-density
DSSD	Double-sided single-density
DSW	Device status word
DTE	Data terminal equipment
DTL	Diode transistor logic
DTMF	Dual tone multi-frequency
DUV	Data under voice
DVM	Digital voltmeter
E	Earth
E^{2}ROM	Electrically erasable read only memory
E^{2}PROM	Electrically erasable programmable read only memory
EAN	European article number
EAM	Electrical accounting machine
EAROM	Electrically alterable read only memory
EBCD	Extended binary coded decimal
EBCDIC	Extended binary coded decimal interchange code
ECD	Electrochromeric display
ECL	Emitter coupled logic
ECMA	European Computer Manufacturers' Association
EDC	Error detection and correction
EDP	Electronic data processing
EDS	Exchangeable disk storage
EFL	Emitter follower logic
EFTS	Electronic funds transfer system
EEROM	Electrically erasable read only memory
EEPROM	Electrically erasable programmable read only memory
EIA	Electronic Industries Association
EITB	Engineering Industries Training Board
EMI	Electromagnetic interference
EOC	End of conversion
EOD	End of data
EOF	End of file
EOM	End of message
EOT	End of text or end of transmission
EPROM	Erasable programmable read only memory
EPU	Extended processing unit
EROM	Erasable read only memory
EXEC	Executive system
F	Farad
FAMOS	Floating gate avalanche metal oxide semiconductor
FAX	Facsimile
FCS	Frame check sequence
FDC	Floppy disk controller
FDM	Frequency division multiplexing
FET	Field effect transistor
FIFO	First-in first-out
FM	Frequency modulation
FOT	Fibre optic transmission
FPGA	Field programmable gate array
FPLA	Field programmable logic array
FP	Floating point
FPU	Floating point unit
FSK	Frequency shift keying
FSM	Frequency shift modulation
G	Giga ($\times 10^9$)
GDP	Graphic display processor

12 Abbreviations (general)

GIGO	Garbage-in garbage-out
GP	General purpose
GPI	General purpose interface
GPIB	General purpose interface bus
H	Henry
Hex	Hexadecimal
Hz	Hertz
HDLC	High level data link control
HMOS	High-density metal oxide semiconductor
HPIB	Hewlett-Packard interface bus
i.c.	Integrated circuit
i/p	Input
I/O	Input/output
I^{2}L	Integrated injection logic
IAS	Immediate access storage
IBG	Inter-block gap
IBM	International Business Machines
ICE	In-circuit emulation
ICL	International Computers Limited
IDPM	Institute of Data Processing Management
IDX	Integrated digital exchange
IEE	Institution of Electrical Engineers
IEEE	Institution of Electrical and Electronics Engineers
IERE	Institution of Electronic and Radio Engineers
IGFET	Insulated gate field effect transistor
IIL	Integrated injection logic
IOP	Input/output processor
IP	Instruction pointer
IPL	Initial program loader
IR	Instruction register
IR	Index register
ISAM	Indexed sequential access method
ISDN	Integrated services digital network
ISDT	Integrated services digital terminal
ISDX	Integrated services digital exchange
ISL	Integrated Schottky logic
ISO	International Standards Organization
ISPBX	Integrated services private branch exchange
ITeC	Information Technology Centre
JAN	Joint Army/Navy
JCL	Job control language
JFET	Junction gate field effect transistor
JUGFET	Junction gate field effect transistor
k	Kilo ($\times 10^3$)
K	Binary kilo ($2^{10} = 1024$)
KSR	Keyboard send/receive
KWIC	Keyword-in-context
L	Inductor
LAN	Local area network
LCD	Liquid crystal display
LED	Light-emitting diode
LF	Line feed
LIFO	Last-in first-out
LOC	Loop on-line control
LP	Line printer
LPM	Lines per minute
LPS	Low-power Schottky
LRC	Longitudinal redundancy check
LRL	Logical record length

Abbreviations (general)

LRU	Last recently used
LS	Low-power Schottky
LSB	Least significant bit
LSD	Least significant digit
LSI	Large scale integration
m	Milli ($\times 10^{-3}$)
M	Mega ($\times 10^6$)
MAP	Macro-arithmetic processor
MAP	Microprocessor Applications Project
MAPCON	Microprocessor Applications Project Consultants
MAR	Memory address register
MBR	Memory buffer register
MCP	Message control program
MCU	Microcomputer unit
MFM	Modified frequency modulation
MICR	Magnetic ink character recognition
MIDI	Musical instrument digital interface
MIS	Metal insulator silicon
MIS	Management information system
MMU	Memory management unit
MNOS	Metal nitride oxide semiconductor
MODEM	Modulator/demodulator
MON	Monitor
MOS	Metal oxide semiconductor
MOSFET	Metal oxide semiconductor field effect transistor
MPU	Microprocessor unit
MPX	Multiplex
MROM	Mask programmed read only memory
MSB	Most significant bit
MSD	Most significant digit
MSI	Medium scale integration
MTBF	Mean time between failure
MTF	Mean time to failure
MTTF	Mean time to failure
MUX	Multiplexer
n	Nano ($\times 10^{-9}$)
NaN	Not a number
NBV	Non-bus vectored
NC	Numerical control
NCC	National Computing Centre
NDP	Numerical data processor
NDRO	Non-destructive readout
NLQ	Near-letter quality
NMOS	N-channel metal oxide semiconductor
NOP	No operation
NRZ	Non return to zero
NRZI	Non return zero invert
NTSC	National Television Systems Committee
NUA	Network user address
NUI	Network user identification
NVM	Non-volatile memory
o/p	Output
OCR	Optical character reader
OCR	Optical character recognition
OEM	Original equipment manufacturer
OIS	Office information systems
OMR	Optical mark reading
OP	Operation
OPCODE	Operation code

14 Abbreviations (general)

OSI	Open systems interconnection
p	Pico ($\times 10^{-12}$)
p.c.	Printed circuit
p.c.b.	Printed circuit board
PABX	Private automatic branch exchange
PAD	Packet assembler/disassembler
PAL	Programmed array logic
PAM	Pulse amplitude modulation
PBX	Private branch exchange
PC	Personal computer
PC	Program counter
PCIO	Program controlled input/output
PCM	Pulse code modulation
PCS	Process control system
PDN	Private data network
PERT	Program evaluation and review technique
PFR	Power failure restart
PIA	Peripheral interface adaptor
PIC	Position independent code
PIC	Program interrupt control
PID	Process identification number
PIO	Programmable input/output
PIPO	Parallel input/parallel output
PISO	Parallel input/serial output
PIT	Programmable interval timer
PLA	Programmable logic array
PLL	Phase-locked loop
PM	Phase modulation
PMOS	P-channel metal oxide semiconductor
POS	Point of sale
POST	Point of sale terminal
PPI	Programmable parallel interface
PPM	Pulse position modulation
PRF	Pulse repetition frequency
PROM	Programmable read only memory
PRT	Program reference table
PSE	Packet switch exchange
PSI	Programmable serial interface
PSN	Packet switched network
PSS	Packet switch stream
PSTN	Public subscriber telephone network
PSW	Processor status word
PTP	Paper tape punch
PTR	Program tape reader
PWB	Printed wiring board
PWM	Pulse width modulation
Q	Transistor
QAM	Quadrature amplitude modulation
QISAM	Queued indexed sequential access method
QUIP	Quad in-line package
R	Resistor
R/W	Read/write
RALU	Register arithmetic logic unit
RAM	Random access memory
RB	Return to bias
RCTL	Resistor capacitor transistor logic
RDSR	Receiver data service request
REC	Rectifier
RJE	Remote job entry

Abbreviations (general) 15

RND	Random
ROM	Read only memory
RPG	Report program generator
RPROM	Reprogrammable read only memory
RTBM	Real-time bit mapping
RTC	Real-time clock
RTE	Real-time executive
RTL	Resistor transistor logic
RX	Receiver
RZ	Return to zero
S/H	Sample and hold
S/N	Signal-to-noise
SA	Signature analysis
SAR	Successive approximation register
SBC	Single-board computer
SC	Short-circuit
SC	Start conversion
SCCD	Surface channel charge-coupled device
SCRN	Screen
SDLC	Synchronous data link control
SDR	Statistical data recorder
SEQ	Sequential
SI	International System
SID	Serial interface device
SID	Symbolic interactive debugger
SIO	Serial input/output
SIPO	Serial input/parallel output
SISO	Serial input/serial output
SLIC	Subscriber line interface circuit
SMC	Surface mounting component
SMD	Storage module disk
SMD	Surface mounting device
SME	Society of Manufacturing Engineers
SMT	Surface mounting technology
SMT	Systems management terminal
SOC	Start of conversion
SOI	Silicon on insulator
SOS	Silicon on sapphire
SP	Stack pointer
SR	Service request
SRAM	Static random access memory
SSDA	Synchronous serial data adaptor
SSDD	Single-sided double-density
SSI	Small scale integration
SSSD	Single-sided single-density
SUB	Subroutine
SYN	Synchronizing
SYGEN	System generation
SYS	System
SYSGEN	System generation
SYSLOG	System log
SYSOP	System operator
TC	Terminal controller
TD	Transmitter distributor
TDM	Time division multiplexing
TDSR	Transmitter data service request
TP	Test point
TPA	Transient program area
TR	Track

16 Abbreviations (general)

TR	Transistor
TRL	Transistor resistor logic
TSL	Tri-state logic
TSS	Time-shared system
TTL	Transistor-transistor logic
TTY	Teletype
TV	Television
TX	Transmitter
UART	Universal asynchronous receiver/transmitter
UBC	Universal block channel
ULA	Uncommitted logic array
UPC	Universal peripheral controller
UPC	Universal product code
UPS	Uninterruptible power supply
USACII	United States standard code for information interchange
USART	Universal asynchronous receiver/transmitter
USRT	Universal synchronous receiver/transmitter
V	Volt
VAB	Voice answer back
VDG	Video display generator
VDI	Virtual device interface
VDP	Video display processor
VDT	Video display terminal
VDU	Visual display unit
VIA	Versatile interface adaptor
VIC	Video interface chip
VIP	Visual information processor
VLSI	Very large scale integration
VM	Virtual memory
VMA	Valid memory address
VMOS	V-groove (vertical) metal oxide semiconductor
VMPU	Virtual memory processing unit
VPA	Valid peripheral address
VRAM	Video random access memory
VRC	Vertical redundancy check
VRC	Visual record computer
VS	Virtual storage
W	Watt
WAN	Wide area network
WATS	Wide area telephone service
WCC	Wild card character
WCS	Writable control store
WIMP	Window icon mouse pull-down
WP	Word processor
WPM	Words per minute
WS	Working store
X	Crystal
XTAL	Crystal
Y	Crystal
Z	Impedance
μ	Micro ($\times 10^{-6}$)
μC	Microcontroller
μP	Microprocessor
Ω	Ohm

Abbreviations commonly used in pin connection data, etc.

a	Anode (diode)
A	General data input (binary weight = 1)
An	Address bus (binary weight = 2^n)
ADn	Multiplexed address/data bus (binary weight = 2^n)
ADRn	Address lines (Intel Multibus)
ALE	Address latch enable (output from CPU)
ARDY	Peripheral port A ready (output from PIO)
ASTB	Strobe pulse for port A (input to PIO)
ATN	Attention (IEEE-488)
b	Base (bipolar transistor)
B	General data input (binary weight = 2)
B	Blue (output to RGB monitor)
BCLK	Bus clock (Intel Multibus)
BCRDY	Ready (IBM PC expansion bus)
BERR	Bus error (input to CPU)
BG	Bus grant (output from CPU)
BGACK	Bus grant acknowledge (input to CPU)
BHEN	Byte high enable (Intel Multibus)
BPRN	Bus priority input (Intel Multibus)
BPRO	Bus priority output (Intel Multibus)
BR	Bus request (input to CPU)
BRDY	Peripheral port B ready (output from PIO)
BREQ	Bus request (Intel Multibus)
BSTB	Strobe pulse for port B (input to PIO)
BUSEN	Bus enable (control input to bus transceiver)
c	Collector (bipolar transistor)
C	General data input (binary weight = 4)
C	Carry (output from hardware adder)
Cin	Carry input
Cout	Carry output
Com.	Common (0V)
C/D	Control/data select input
CAn	Peripheral control line (port A)
CAS	Column address strobe (dynamic RAM)
CBn	Peripheral control line (port B)
CBRQ	Common bus request (Intel Multibus)
CCLK	Constant clock (Intel Multibus)
CE	Chip enable
CHCK	Channel check (IBM PC expansion bus)
CI	Carry input
CK	Clock input
CLK	Clock input
CLR	Clear input
CO	Carry output
CS	Chip select
CTS	Clear to send (RS-232C)
CY	Carry output
d	Drain (FET)
D	Data input for bistable latch
D	General data input (binary weight = 8)
Di	Data input (RAM)
Din	Data input (RAM)
Dn	Data bus (binary weight = 2^n)
Do	Data output (RAM)

18 Abbreviations commonly used in pin connection data, etc.

Dout	Data output (RAM)
DACKn	DMA acknowledge (IBM PC expansion bus)
DATn	Data line (Intel Multibus)
DAV	Data valid (IEEE-488)
DCD	Data carrier detect (RS-232C)
DIS	Disable input for tri-state devices
DTR	Data terminal ready (RS-232C)
DRQn	DMA request (IBM PC expansion bus)
e	Emitter (bipolar transistor)
E	Earth
EN	Enable input
EOI	End or identify (IEEE-488)
FCn	Function code (output from CPU)
FG	Frame ground (RS-232C)
g	Gate (FET)
G	Enable input for tri-state devices
G	Green (output to RGB monitor)
Gnd	Ground, common, 0 V
GND	Ground, common, 0 V
I/O	Input/output mode control
I/On	Input/output (RAM, binary weight = 2^n)
IEI	Interrupt enable input (PIO)
IEO	Interrupt enable output (PIO)
IFC	Interface clear (IEEE-488)
IIOR	I/O read (IBM PC expansion bus)
IIOW	I/O write (IBM PC expansion bus)
IMW	Memory write (IBM PC expansion bus)
IMR	Memory read (IBM PC expansion bus)
INH	Inhibit
INHn	Inhibit (Intel Multibus)
INIT	Initialize (Centronics printer bus, Intel Multibus)
INT	Interrupt request (input to CPU)
INTn	Interrupt request (Intel Multibus)
INTA	Interrupt acknowledge (output from CPU)
INTR	Maskable interrupt (input to CPU)
IOn	Input/output (RAM, binary weight = 2^n)
IORC	Input/output read (Intel Multibus)
IORQ	Input/output request (output from CPU)
IOWC	Input/output write (Intel Multibus)
IPLn	Interrupt priority level (n denotes bit significance, e.g. 0 = LSB)
IRQ	Interrupt request (input to CPU)
J	Data input for J-K bistable
k	Cathode (diode)
K	Data input for J-K bistable
LDS	Lower data strobe output
LE	Latch enable input
LSB	Least significant bit
LT	Lamp test input for display driver
M1	First machine cycle (output from Z80-type CPU)
MR	Master reset input
MRDC	Memory read (Intel Multibus)
MREQ	Memory request (output from CPU)
MRQ	Memory request (output from CPU)
MSB	Most significant bit
MWTC	Memory write (Intel Multibus)
n.c.	Not connected
N.C.	Not connected
NDAC	Not data accepted (IEEE-488)

Abbreviations commonly used in pin connection data, etc. 19

NMI	Non-maskable interrupt input
NRFD	Not ready for data (IEEE-488)
o.c.	Open circuit or open collector
O/C	Open circuit
On	Output (ROM, binary weight = 2^n)
OE	Output enable (input to support/memory device)
OEN	Output enable (input to support/memory device)
OV	Overflow
OVF	Overflow
P/S	Parallel/serial shift register mode control
PAn	Peripheral data line (port A)
PBn	Peripheral data line (port B)
PE	Paper end (Centronics printer bus)
PG	Protective ground (RS-232C)
PGM	Program control input (EPROM)
PH	Phase input for LCD display
PR	Preset input
Q	General output for bistable, latch, counter, and shift register
Qn	Output from latch, counter, or shift register (binary weight = 2^n)
R	Red (output to RGB monitor)
R	Reset input for R-S bistable
R/B	Ready/not busy (EEPROM program control output)
R/W	Read/write select output
RxD	Receive data (RS-232C)
RC	Receive clock (RS-232C)
RD	Receive data (RS-232C) or read (output from CPU)
REN	Remote enable (IEEE-488)
RES	Reset (input to CPU and support devices)
RFSH	Refresh (output from CPU)
RSn	Register select (n denotes register)
RTS	Request to send (RS-232C)
s	Source (FET)
s.c.	Short circuit
S/C	Short circuit
S	Set input for R-S bistable or sum (output of hardware adder)
Sin	Serial input of shift register
Sout	Serial output of shift register
SDL	Serial data input (left shift)
SDR	Serial data input (right shift)
SEL	Select input for multiplexer/demultiplexer
SG	Signal ground (RS-232C)
SLCT	Select (Centronics printer bus)
ST	Strobe input
SR	Synchronous reset input/output
T	Trigger input
TxD	Transmit data (RS-232C)
TC	Transmit clock (RS-232C)
TC	Terminal count (IBM PC expansion bus)
TD	Transmit data (RS-232C)
U/D	Up/down select input for counter
UDS	Upper data strobe output
Vbb	Negative supply for RAM device (usually -5 V)
Vcc	TTL positive supply (usually $+5$ V)
Vdd	CMOS positive supply (often $+5$ V)
Vee	ECL negative supply (usually -5 V)
Vi	Analogue input voltage (A to D converter)

Vin	Analogue input voltage (A to D converter)
Vo	Analogue output voltage (D to A converter)
Vout	Analogue output voltage (D to A converter)
Vpp	Programming voltage (EPROM)
Vss	Voltage source-source (CMOS common 0 V)
VID	Video
VMA	Valid memory address (output from CPU)
VPA	Valid peripheral address (output from CPU)
W	Write output
WE	Write enable
WR	Write output
X	General output for logic gate arrangement
X	Data (input to data selector)
XACK	Transfer acknowcedge (Intel Multibus)
Y	General output for logic gate arrangement
Y	Data (input to data selector)
Y	Luminance (output to PAL encoder)
Z	General output for logic gate arrangement
Z	Data (input to data selector)
0 V	Common, ground
—	Active low
ϕ	Clock (input to CPU)
ϕn	Clock input/output (n denotes phase)
$\varGamma$	Schmitt device
Σ	Sum output of hardware adder

Note: n = 0, 1, 2, 3, etc.

Manufacturers' prefixes for semiconductor devices

AD	Analog Devices
AD	Intersil
AH	National Semiconductor
AM	AMD
AY	General Instrument
C	Intel
CD, CDP	RCA
CP	General Instrument
D	Intel
DG	Siliconix
DM	National Semiconductor
DMPAL	National Semiconductor
DS	National Semiconductor
DS	Signetics
DS	Texas Instruments
DP	AMD
DP	National Semiconductor
EF	Thomson/EFCIS
F	Fairchild
F	Ferranti
G	GTE
H	SGS
HCMP	Hughes
HD	Hitachi
HEF	Mullard
HEF	Signetics

Manufacturers' prefixes for semiconductor devices

HM	Hitachi
HN	Hitachi
I	Intel
ICL	Intersil
ICM	Intersil
IM	Intersil
INS	National Semiconductor
KMM	Texas Instruments
LF	National Semiconductor
LM	National Semiconductor
LM	Signetics
LM	Texas Instruments
LS	Texas Instruments
NM	National Semiconductor
M	Mitsubishi
MAB	Mullard
MBL	Fujitsu
MC	Motorola
MC	Signetics
MC	Texas Instruments
MJ	Plessey
MK	Mostek
ML	Plessey
MM	National Semiconductor
MN	Plessey
MP	MPS
MSM	OKI
MV	Plessey
N	Signetics
NE	Signetics
NJ	Plessey
NS	National Semiconductor
NSC	National Semiconductor
P	AMD
P	Intel
PC	Signetics
PCF	Signetics
PIC	Plessey
R	Rockwell
R	Raytheon
RAY	Raytheon
RC	Raytheon
S	American Microsystems
SAA	Signetics
SCB	Signetics
SCN	Signetics
SCP	Solid State Scientific
SE	Signetics
SL	Plessey
SN	Texas Instruments
SP	Plessey
SY	Synertek
TAB	Plessey
TBP	Texas Instruments
TC	Toshiba
TCA	Signetics
TCM	Texas Instruments
TDA	Signetics
TEA	Signetics

TIC	Texas Instruments
TIL	Texas Instruments
TIM	Texas Instruments
TIP	Texas Instruments
TL	Texas Instruments
TLC	Texas Instruments
TMM	Toshiba
TMP	Texas Instruments
TMS	Texas Instruments
UA	Signetics
UA	Texas Instruments
UCN	Sprague
UDN	Sprague
UDN	Texas Instruments
UGN	Sprague
ULN	Signetics
ULN	Sprague
ULN	Texas Instruments
UPB	NEC
UPD	NEC
X	Xicor
Z	Zilog
Z	SGS
ZN	Ferranti
μPD	NEC

Integrated circuit technologies

A variety of different semiconductor technologies are currently employed in the manufacture of integrated circuits. These technologies are instrumental in governing the operational characteristics of devices and an awareness of their essential differences can be useful in selecting devices for use in a particular hardware configuration.

P-channel metal oxide semiconductor (PMOS)
PMOS devices use enhancement-mode p-channel MOS transistors to form gates. PMOS devices employ positive charge carriers which are known as 'holes'. All of the first generation of microprocessors were based upon PMOS technology (the process was originally employed in preference to NMOS by virtue of its relative freedom from contamination).

PMOS devices have typical densities of around 20 000 devices per chip but are relatively slow in operation (a typical PMOS microprocessor is capable of executing around 500 000 instructions per second).

N-channel metal oxide semiconductor (NMOS)
NMOS devices use n-channel MOS transistors, in which electrons rather than holes are employed as charge carriers. NMOS devices provide excellent densities (over 100 000 devices per chip) and operate at acceptably high speeds (a typical NMOS microprocessor is capable of executing 1 million instructions per second).

A number of variants of NMOS are in common use. These include:

(a) Double diffused metal oxide semiconductor (DMOS)
(b) High-density metal oxide semiconductor (HMOS)
(c) V-groove metal oxide semiconductor (VMOS)

Complementary metal oxide semiconductor (CMOS)

CMOS employs both n-channel and p-channel devices and its performance is therefore something of a compromise between the two technologies. CMOS provides densities of typically around 50 000 devices per chip but is inherently slower than NMOS.

Despite its speed and density limitations, CMOS offers several important advantages over its rival technologies. It requires very little power and operates over a very wide range of supply voltages. It also offers excellent noise immunity.

Readers should note that the power consumption of a CMOS device is directly related to the speed at which the device is operating. Furthermore, in a 'standby' condition, such devices consume negligible power.

Bipolar

Bipolar technology is based on conventional NPN junction devices and exists in several forms of which the most common are:

(a) transistor-transistor logic (TTL)
(b) low-power Schottky transistor-transistor logic (LS-TTL)
(c) emitter-coupled logic (ECL)
(d) integrated injection logic (I^2L)

TTL devices are extremely fast in operation (equivalent to 10 million instructions per second) but consume appreciable power. This makes them unsuitable for high-density applications (such as complete microprocessors).

ECL devices offer the highest switching speed of all but they demand so much power that only small scale integration is possible. For this reason ECL devices are reserved for such specialized applications as HF and VHF measuring equipment and the synthesis of VHF signals.

Integrated injection logic (I^2L) provides moderately high speed operation coupled with low power consumption. I^2L devices provide moderately good packing densities and have thus proved popular in the implementation of 'bit-slice' devices. I^2L devices are also popularly used in pocket calculators.

Charge-coupled devices (CCD)

As the name implies, charge-coupled devices rely upon charge rather than current carriers for their operation. CCD comprise a large matrix of individual capacitors formed by deposition of aluminium on silicon oxide. In order to replace the charge that must eventually leak away from an elementary cell, cells are periodically refreshed by the regular shifting of charges from one cell to the next in a constant circulating manner. Densities are excellent but the technology is still relatively new and its cost-effectiveness has yet to be proved.

Scale of integration

The density of integration of a device is popularly expressed in terms of its relative scale of operation. The following terms are commonly used but their meaning (in terms of the number of equivalent logic gates) is open to some variation:

(a) Small scale integration (SSI)
    Logic gate equivalent: 1 to 10
    Typical examples: TTL logic gates

24 Basic logic gates

(b) Medium scale integration (MSI)
 Logic gate equivalent: 10 to 100
 Typical examples: Bipolar memories
(c) Large scale integration (LSI)
 Logic gate equivalent: 100 to 1000
 Typical examples: Programmed logic arrays
(d) Very large scale integration (VLSI)
 Logic gate equivalent: 1000 to 10 000
 Typical examples: Most common microprocessors
(e) Super large scale integration (SLSI)
 Logic gate equivalent: 10 000 to 100 000
 Typical examples: NMOS dynamic RAM

Basic logic gates

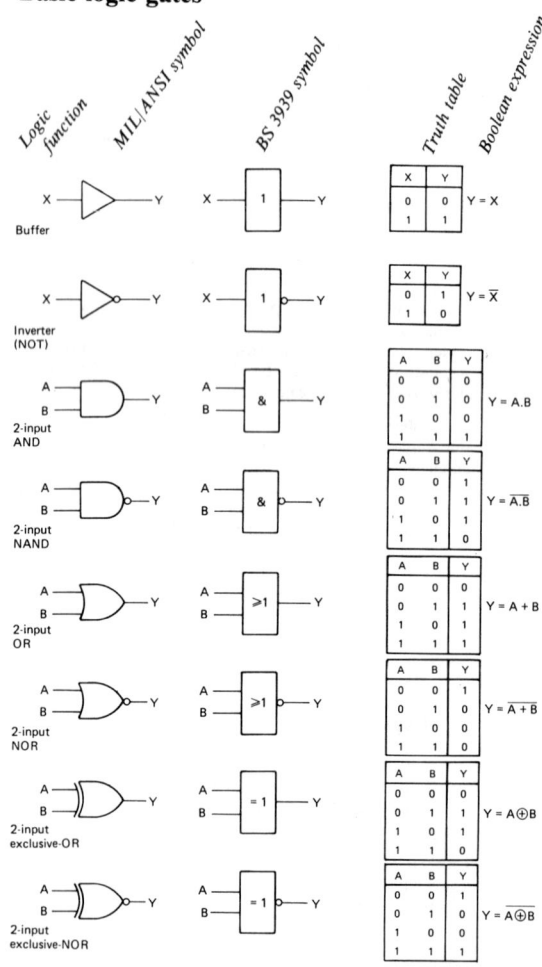

Logic circuit equivalents

The following logic circuit equivalents are useful when it is necessary to minimize the number of logic gates in a given arrangement or when a restriction is placed on the types of gate available. It should be noted that, while the logical functions will be identical, the electrical performance may be different. This is particularly true in the case of propagation delay.

Positive and negative logic equivalents

Positive Logic		Negative Logic
AND	=	OR
OR	=	AND
NAND	=	NOR
NOR	=	NAND

Positive Logic: Logic 1 = High, Logic 0 = Low
Negative Logic: Logic 1 = Low,   Logic 0 = High

Mixed logic equivalents

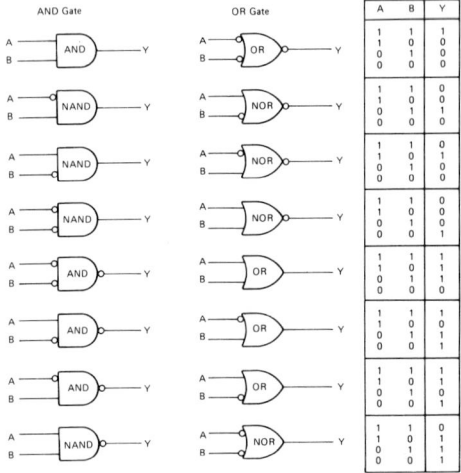

Typical CMOS and TTL gate circuits

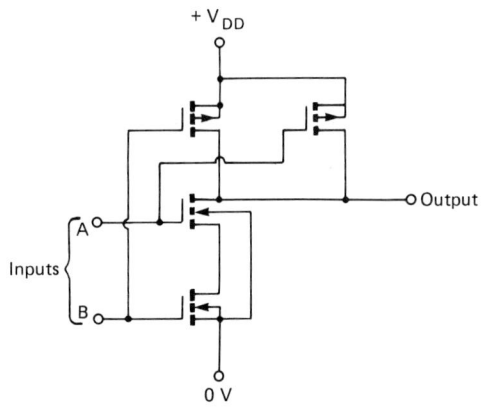

CMOS 2-input NAND

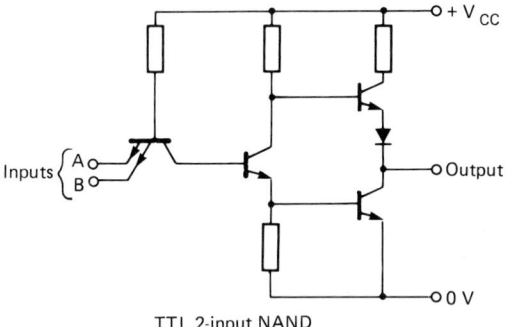

TTL 2-input NAND

TTL and CMOS device coding

TTL
The most common family of TTL devices is the 74-series, in which the device code is prefixed by the number 74. In addition, manufacturers may introduce their own identifying prefix letters. Common examples are:

N	=	Signetics
CD	=	RCA
DM	=	National Semiconductor
MC	=	Motorola
MM	=	National Semiconductor
PC	=	Signetics
SN	=	Texas Instruments

28 TTL and CMOS device coding

Additional letters may be inserted within the device coding to indicate the sub-family to which the device belongs:

C	=	CMOS version of the TTL device
F	=	fast
H	=	high speed
L	=	low power
S	=	Schottky
HC	=	high speed CMOS version (with CMOS-compatible inputs)
HCT	=	high speed CMOS version (with TTL-compatible inputs)
LS	=	low power Schottky
ALS	=	advanced low power Schottky

A suffix letter may also be added to denote the type of package. The most common is N, which describes the conventional plastic dual-in-line (DIL) package.

Examples
1. SN7400N is a quad 2-input NAND gate manufactured by Texas Instruments and supplied in a plastic DIL package.
2. MM74HC32N is a quad 2-input NOR gate using high speed CMOS technology which is speed, function, and pinout compatible with 74LS-series logic. The device is supplied in a plastic DIL package and is manufactured by National Semiconductor.
3. N74LS373N is a low power Schottky octal tri-state latch. The device is manufactured by Signetics and supplied in a plastic DIL package.

CMOS

The most common CMOS families are the 4000 and 4500 series of devices. Devices are usually coded with a suffix letter which indicates the series to which the device belongs. The original A-series is now largely obsolete and has been replaced by the pin and function compatible B-series. These devices feature buffered outputs. Unbuffered B-series devices are also available and these are coded with the suffix letters UB.

As with TTL devices, a manufacturer's prefix may be added. Common examples are:

CD	=	National Semiconductor
CD	=	RCA
HEF	=	Signetics
MC1	=	Motorola

A final suffix letter may also be added in order to specify the type of package. Manufacturers' literature should be consulted for further information.

Examples
1. CD4002BF is a quad 4-input NOR gate manufactured by RCA. The device is a buffered B-series device and supplied in a ceramic DIL package.
2. HEF4069UBP is a hex inverter produced by Signetics. The device features an unbuffered output and is supplied in a plastic DIL package.

Abbreviations

The following abbreviations are used in the list of TTL and CMOS devices:

h.v.	=	high voltage
inv.	=	inverting
o.c.	=	open collector
o.d.	=	open drain
str.	=	strobed
ALU	=	arithmetic logic unit
BCD	=	binary coded decimal
FIFO	=	first-in first-out
GPIB	=	general purpose instrument bus
PIPO	=	parallel-input parallel-output
PISO	=	parallel-input serial-output
SIPO	=	serial-input parallel-output
SISO	=	serial-input serial-output

74 series

00	Quad 2-input NAND
01	Quad 2-input o.c. NOR
02	Quad 2-input NOR
03	Quad 2-input o.c. NAND
04	Hex inverter
05	Hex o.c. inverter
06	Hex o.c. h.v. inverter
07	Hex o.c. h.v. buffer
08	Quad 2-input AND
09	Quad 2-input o.c. AND
10	Triple 3-input NAND
11	Triple 3-input AND
12	Triple 3-input o.c. NAND
13	Dual 4-input Sch. NAND
14	Hex Sch. inverter
15	Triple 3-input o.c. AND
16	Hex o.c. h.v. inverter
17	Hex o.c. h.v. buffer
20	Dual 4-input NAND
21	Dual 4-input AND
22	Dual 4-input o.c. NAND
23	Dual 4-input str. NOR
25	Dual 4-input str. NOR
26	Quad 2-input o.c. NAND
27	Triple 3-input NOR
28	Quad 2-input NOR buffer
30	Single 8-input NAND
32	Quad 2-input OR
33	Quad 2-input o.c. NOR buffer
37	Quad 2-input NAND buffer
38	Quad 2-input o.c. NAND buffer
40	Dual 4-input NAND buffer
42	BCD to decimal decoder
43	Excess 3 to decimal decoder
44	Gray to decimal decoder
45	BCD to decimal o.c. h.v. decoder
46	BCD to 7-segment o.c. h.v. decoder
47	BCD to 7-segment o.c. h.v. decoder
48	BCD to 7-segment decoder

49	BCD to 7-segment decoder
50	Dual AND/OR/invert
51	Dual AND/OR/invert
52	Single AND/OR
53	Single AND/OR/invert
54	Single AND/OR/invert
55	Single AND/OR/invert
60	Dual 4-input expander
61	Triple 3-input expander
62	Single AND/OR expander
63	Hex current-sensing interface
64	Single AND/OR/invert
65	Single AND/OR/invert
70	Single J-K bistable
71	Single J-K bistable
72	Single J-K bistable
73	Dual J-K bistable
74	Dual D-type bistable
75	Quad bistable latch
76	Dual J-K bistable
77	4-bit bistable latch
78	Dual J-K bistable
80	2-bit full adder
81	16-bit RAM
82	2-bit full adder
83	4-bit full adder
84	16-bit RAM
85	4-bit comparator
86	Quad 2-input exclusive-OR
87	4-bit complementor
88	256-bit ROM
89	64-bit RAM
90	Decade counter
91	8-bit shift register
92	Divide-by-twelve counter
93	4-bit binary counter
94	4-bit shift register
95	4-bit shift register
96	5-bit shift register
97	6-bit binary rate multiplier
98	4-bit data selector
99	4-bit bi-directional shift register
100	Dual 4-bit latch
101	Single J-K bistable
102	Single J-K bistable
103	Dual J-K bistable
104	Single J-K bistable
105	Single J-K bistable
106	Dual J-K bistable
107	Dual J-K bistable
108	Dual J-K bistable
109	Dual J-K bistable
110	Single J-K bistable
111	Dual J-K bistable
112	Dual J-K bistable
113	Dual J-K bistable
114	Dual J-K bistable
116	Dual 4-bit latch
118	Hex R-S bistable latch

119	Hex R-S latch
120	Dual pulse synchronizer
121	Monostable
122	Retriggerable monostable
123	Dual retriggerable monostable
124	Dual voltage controlled oscillator
125	Quad tri-state buffer
126	Quad tri-state buffer
128	Quad 2-input NOR line driver
132	Quad 2-input Sch. NAND
133	Single 13-input NAND
134	Single 12-input tri-state NAND
135	Quad exclusive-OR/NOR
136	Quad 2-input exclusive-OR
137	3 to 8-line decoder
138	3 to 8-line decoder
139	Dual 2 to 4-line decoder
140	Dual 4-input NAND
141	BCD to decimal decoder
142	4-bit counter/latch/decoder/driver
143	4-bit counter/latch/decoder/driver
144	4-bit counter/latch/decoder/driver
145	BCD to decimal converter
147	Decimal to 4-bit BCD encoder
148	8 to 3-line octal encoder
150	1-of-16 data selector/multiplexer
151	1-of-8 data selector/multiplexer
152	1-of-8 data selector/multiplexer
153	Dual 4 to 1-line data selector/multiplexer
154	4 to 16-line decoder
155	Dual 2 to 4-line decoder
156	Dual 2 to 4-line o.c. decoder
157	Quad 2 to 1-line data selector
158	Quad 2 to 1-line data selector
159	4 to 16-line o.c. decoder
160	4-bit counter
161	4-bit counter
162	4-bit counter
163	4-bit counter
164	8-bit SIPO shift register
165	8-bit PISO shift register
166	8-bit PISO/SISO shift reg.
167	Decade synchronous rate multiplier
168	4-bit up/down synchronous decade counter
169	4-bit up/down synchronous binary counter
170	4-by-4 o.c. register file
172	16-bit tri-state register file
173	4-bit D-type tri-state register
174	Hex D-type bistable
175	Quad D-type bistable
176	Presettable decade counter/latch
177	Presettable binary counter/latch
178	4-bit universal shift register
179	4-bit universal shift register
180	9-bit parity generator/checker
181	ALU
182	Look-ahead carry generator
183	Dual full adder
184	BCD to binary code converter

32 74 series

185	Binary to BCD code converter
186	512-bit o.c. PROM
187	1K-bit o.c. ROM
188	256-bit o.c. PROM
189	64-bit RAM
190	BCD synchronous up/down counter
191	Binary synchronous up/down counter
192	BCD synchronous dual clock up/down counter
193	Binary synchronous dual clock up/down counter
194	4-bit bidirectional universal shift register
195	4-bit parallel-access shift register
196	Presettable decade counter/latch
197	Presettable binary counter/latch
198	8-bit bidirectional universal shift register
199	8-bit bidirectional universal shift register
200	256-bit tri-state RAM
201	256-bit tri-state RAM
202	256-bit tri-state RAM
207	1K-bit RAM
208	1K-bit tri-state RAM
214	1K-bit tri-state RAM
215	1K-bit tri-state RAM
221	Dual monostable
225	Asynchronous FIFO memory
226	4-bit parallel latched bus transceiver
240	Octal tri-state inv. buffer/line driver/receiver
241	Octal tri-state non-inv. buffer/line driver/receiver
242	Quad tri-state inv. bus transceiver
243	Quad tri-state non-inv. bus transceiver
244	Octal tri-state non-inv. buffer/line driver/receiver
245	Octal tri-state non-inv. bus transceiver
246	BCD to 7-segment o.c. h.v. decoder
247	BCD to 7-segment o.c. h.v. decoder
248	BCD to 7-segment decoder
249	BCD to 7-segment o.c. decoder
251	8 to 1-line tri-state data selector
253	Dual 4 to 1-line tri-state data selector
256	Dual 4-bit latch
257	Quad 2-input tri-state non-inv. multiplexer
258	Quad 2-input tri-state inv. multiplexer
259	8-bit addressable latch
260	Dual 5-input NOR
261	2 by 4-bit parallel binary multiplier
265	Quad complementary output generator
266	Quad 2-input o.c. exclusive-NOR
270	2K-bit ROM
271	2K-bit ROM
273	Octal D-type bistable
274	4 by 4-bit tri-state binary multiplier
275	7-bit-slice Wallace tree
276	Quad J-K bistable
278	4-bit cascadable priority register
279	Quad R-S latch
280	9-bit parity generator/checker
281	4-bit parallel binary accumulator
283	4-bit binary full adder
284	4 by 4-bit parallel binary multiplier
285	4 by 4-bit parallel binary multiplier
287	1K-bit tri-state PROM

74 series

288	256-bit tri-state PROM
289	64-bit o.c. RAM
290	Decade counter
293	4-bit binary counter
295	4-bit bi-directional universal shift register
298	Quad 2-input multiplexer
299	8-bit bidirectional universal shift register
300	256-bit o.c. RAM
301	256-bit o.c. RAM
302	256-bit o.c. RAM
314	1K-bit o.c. RAM
315	1K-bit o.c. RAM
320	Crystal controlled oscillator
321	Crystal controlled oscillator
323	8-bit bidirectional universal shift register
324	Voltage controlled oscillator
325	Dual voltage controlled oscillator
326	Dual voltage controlled oscillator
327	Dual voltage controlled oscillator
348	8 to 3-line tri-state priority encoder
351	Dual 8 to 1-line tri-state data selector
352	Dual 4 to 1-line inv. data selector
353	Dual 4 to 1-line tri-state inv. data selector
354	8 to 1-line data selector
356	8 to 1-line data selector
362	Four-phase clock generator
363	Octal tri-state D-type latch
364	Octal tri-state D-type latch
365	Hex tri-state non-inv. buffer
366	Hex tri-state inv. buffer
367	Hex tri-state non-inv. buffer
368	Hex tri-state inv. buffer
370	2K-bit tri-state ROM
371	2K-bit tri-state ROM
373	Octal tri-state D-type latch
374	Octal tri-state D-type bistable
375	4-bit bistable latch
376	Quad J-K bistable
377	Octal D-type bistable
378	Hex D-type bistable
379	Quad D-type bistable
381	4-bit ALU
386	Quad 2-input exclusive-OR
387	1K-bit o.c. PROM
390	Dual decade counter
393	Dual 4-bit binary counter
395	4-bit tri-state universal shift register
398	Quad 2-input multiplexer
399	Quad 2-input multiplexer
412	8-bit tri-state buffered latch
423	Dual monostable
424	Two-phase clock generator
425	Quad tri-state buffer
426	Quad tri-state buffer
428	Bidirectional system controller
438	Bidirectional system controller
442	Quad tri-state bus transceiver
443	Quad tri-state bus transceiver
444	Quad tri-state bus transceiver

446	Quad bus transceiver
449	Quad bus transceiver
470	256 by 8-bit o.c. PROM
471	256 by 8-bit tri-state PROM
472	512 × 8-bit tri-state PROM
473	512 × 8-bit o.c. PROM
474	512 × 8-bit tri-state PROM
475	512 × 8-bit o.c. PROM
476	1K × 4-bit tri-state PROM
477	1K × 4-bit o.c. PROM
481	4-bit slice processor
482	4-bit slice controller
490	Dual decade counter
500	6-bit analogue to digital converter
505	8-bit analogue to digital converter
521	Octal comparator
524	8-bit register comparator
533	Octal tri-state latch
534	Octal tri-state latch
537	1 of 10 tri-state decoder
538	1 of 8 tri-state decoder
539	Dual 1 of 4 tri-state decoder
540	Octal tri-state inv. buffer
541	Octal tri-state non-inv. buffer
543	Octal bidirectional latch
544	Octal bidirectional latch
545	Octal bus transceiver
547	3 to 8-line decoder
548	3 to 8-line decoder
550	Register transceiver
551	Register transceiver
552	8-bit register
557	8 × 8 multiplier
558	8 × 8 multiplier
563	Octal tri-state latch
564	Octal inv. tri-state latch
568	Decade tri-state up/down counter
569	Binary tri-state up/down counter
573	Octal tri-state latch
574	Octal tri-state non-inv. D-type bistable
576	Octal tri-state inv. D-type bistable
579	8-bit up/down counter
580	Octal tri-state inv. latch
582	4-bit BCD ALU
583	4-bit BCD adder
588	Octal transceiver (GPIB compatible)
589	8-bit shift register
590	8-bit binary counter
592	8-bit binary counter
593	8-bit binary counter
594	8-bit shift register
595	8-bit shift register
597	8-bit shift register
598	8-bit shift register
604	Dual octal tri-state latch
605	Dual octal tri-state latch
606	Dual octal tri-state latch
610	Tri-state memory mapper
612	Tri-state memory mapper

74 series

620	Octal tri-state inv. bus transceiver
621	Octal o.c. non-inv. bus transceiver
622	Octal o.c. inv. bus transceiver
623	Octal tri-state non-inv. bus transceiver
624	Voltage controlled oscillator
625	Dual voltage controlled oscillator
626	Dual voltage controlled oscillator
627	Dual voltage controlled oscillator
628	Dual two-phase voltage controlled oscillator
629	Dual oscillator
638	Octal tri-state inverting o.c. bus transceiver
639	Octal tri-state inverting o.c. bus transceiver
640	Octal tri-state inv. bus transceiver
641	Octal o.c. non-inv. bus transceiver
642	Octal o.c. inv. bus transceiver
643	Octal tri-state inv. bus transceiver
644	Octal o.c. non-inv. bus transceiver
645	Octal tri-state non-inv. bus transceiver
646	Octal tri-state bus transceiver
647	Octal o.c. bus transceiver
648	Octal tri-state bus transceiver
649	Octal o.c. bus transceiver
651	Octal bus transceiver
652	Octal bus transceiver
668	Synchronous decade up/down counter
669	Synchronous binary up/down counter
670	4 × 4-bit tri-state register file
673	16-bit SISO shift register
674	16-bit PISO shift register
675	16-bit SIPO shift register
676	16-bit PISO shift register
681	ALU
682	8-bit comparator
683	8-bit o.c. comparator
684	8-bit comparator
685	8-bit o.c. comparator
686	8-bit comparator
687	8-bit o.c. comparator
688	8-bit comparator
689	8-bit o.c. comparator
690	8-bit tri-state decade counter
691	4-bit tri-state binary counter
692	4-bit tri-state decade counter
693	4-bit tri-state binary counter
696	4-bit tri-state decade up/down counter
697	4-bit tri-state binary up/down counter
698	4-bit tri-state decade up/down counter
699	4-bit tri-state binary up/down counter
740	Octal tri-state inv. buffer/driver
741	Octal tri-state non-inv. buffer/driver
744	Octal tri-state non-inv. buffer/driver
748	8 to 3-line encoder
779	8-bit up/down counter
784	8-bit serial multiplier
795	Octal tri-state buffer
796	Octal tri-state buffer
797	Octal tri-state buffer
798	Octal tri-state buffer
804	Hex 2-input NAND line driver

74 series

805	Hex 2-input NOR line driver
808	Hex 2-input AND line driver
832	Hex 2-input OR line driver
848	8 to 3-line tri-state priority encoder
857	8-line multiplexer
873	Dual quad latch
874	Dual quad D-type bistable
876	Dual quad D-type bistable
878	Dual quad D-type bistable
879	Octal inv. D-type bistable
880	Dual quad inv. latch
881	4-bit ALU
882	32-bit look-ahead carry generator
901	Hex inv. buffer (TTL interface)
902	Hex buffer (TTL interface)
903	Hex inv. buffer (MOS interface)
904	Hex buffer (MOS interface)
905	12-bit successive approximation register
906	Buffer o.d.
907	Buffer o.d.
908	Dual h.v. CMOS driver
909	Quad comparator
910	256-bit RAM
911	4-digit 8-segment display controller
912	6-digit 8-segment display controller
914	Hex Sch. trigger
915	7-segment to BCD decoder
917	6-digit hex display controller
918	Dual h.v. CMOS driver
922	16-key keyboard encoder
923	20-key keyboard encoder
925	4-digit counter/7-segment display driver
926	4-digit counter/7-segment display driver
927	4-digit counter/7-segment display driver
928	4-digit counter/7-segment display driver
932	Phase detector
941	Octal tri-state buffer/line receiver/line driver
945	4-digit LCD up/down counter/latch/driver
946	4-digit LCD up/down counter/latch/driver
947	4-digit LCD up/down counter/latch/driver
956	4-digit 17-segment alpha-numeric display decoder/driver
1000	Buffered 00
1002	Buffered 02
1003	Buffered 03
1004	Buffered 04
1005	Buffered 05
1008	Buffered 08
1010	Buffered 10
1011	Buffered 11
1020	Buffered 20
1032	Buffered 32
1034	Hex buffer
1035	Hex buffer
1036	Quad 2-input NOR line driver
1240	Low power 240
1241	Low power 241
1242	Low power 242
1243	Low power 243
1244	Low power 244

1245	Low power 245
1616	16 × 16 multiplier
1620	Low power 620
1621	Low power 621
1622	Low power 622
1623	Low power 623
1638	Low power 638
1639	Low power 639
1640	Low power 640
1641	Low power 641
1642	Low power 642
1643	Low power 643
1644	Low power 644
1645	Low power 645
2620	Octal bus transceiver
2623	Octal bus transceiver
2640	Octal bus transceiver
2645	Octal bus transceiver

74 Series pin connections

00

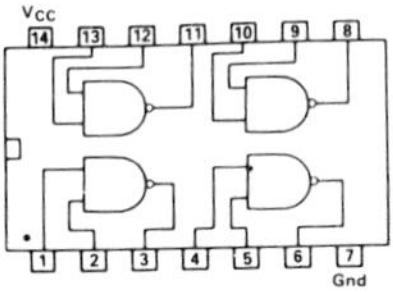

01

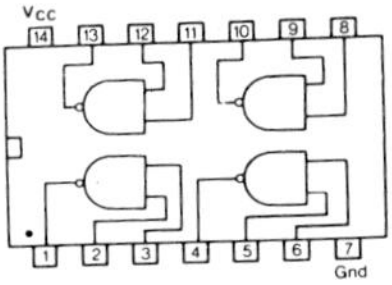

02

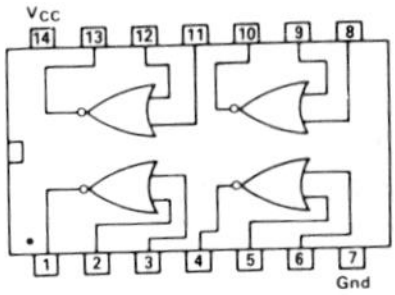

03

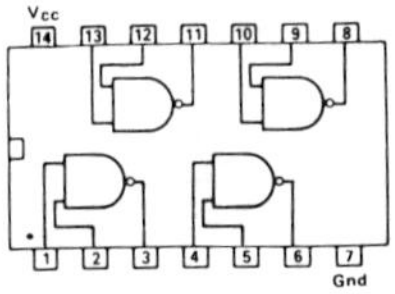

04

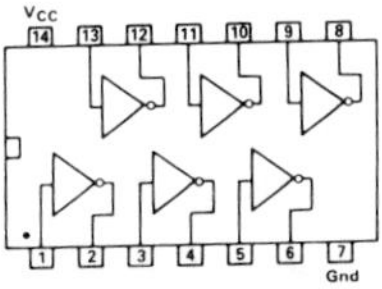

05

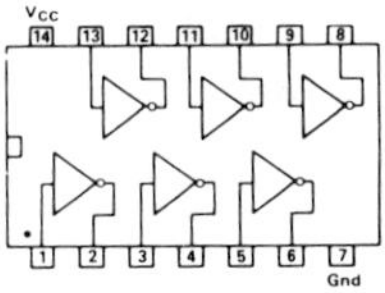

38 74 series

06

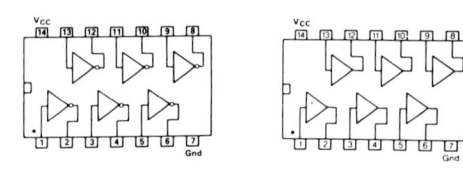

07

08

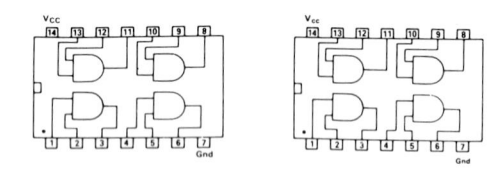

09

10

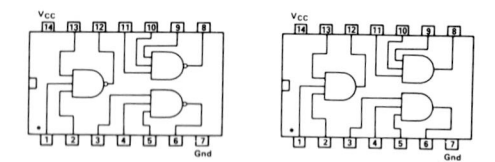

11

13

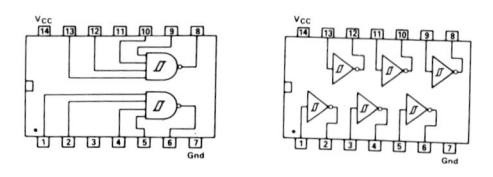

14

15

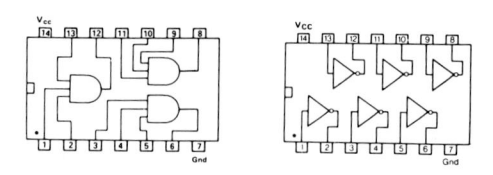

16

74 series 39

20

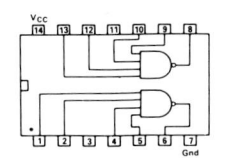

21

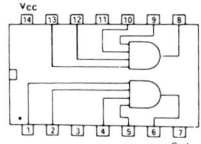

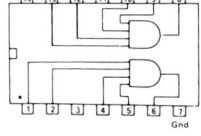

22

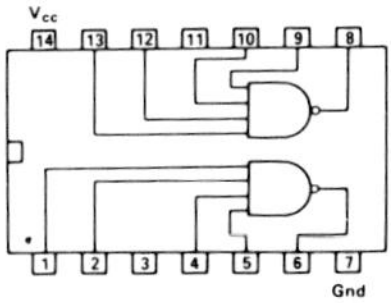

25

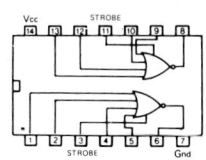

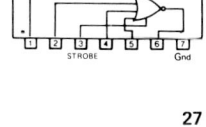

26

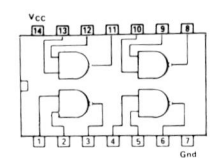

27

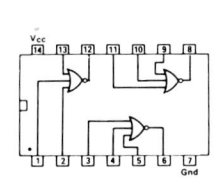

28

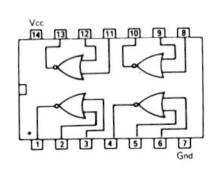

30

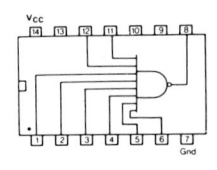

32

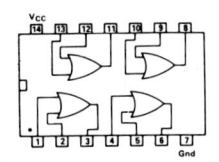

33

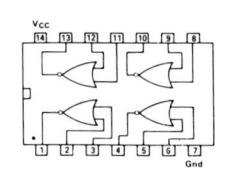

40 74 series

37

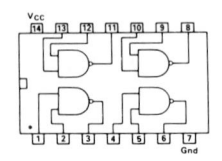

38

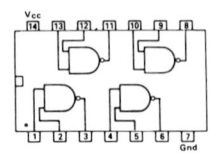

40

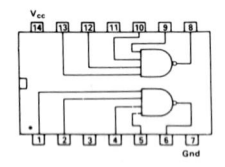

70

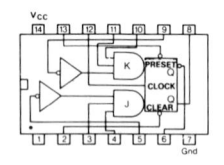

72

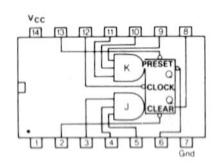

73

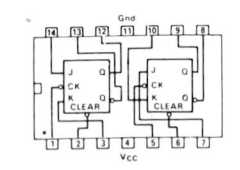

74

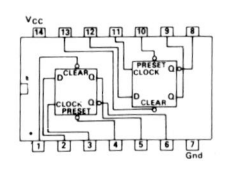

75

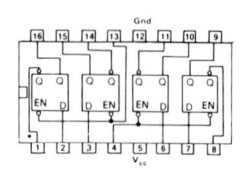

76

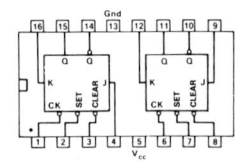

78

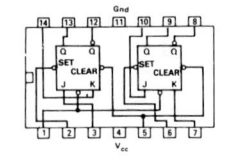

74 series

86

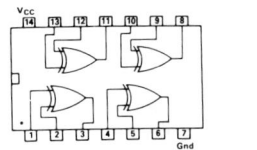

107

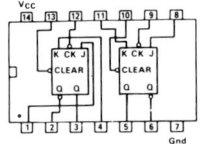

109

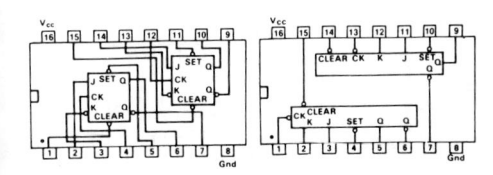

112

113

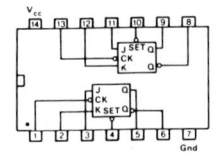

114

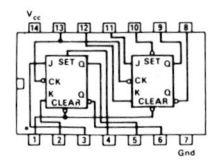

121

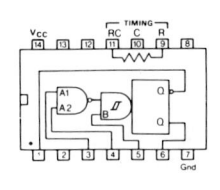

123

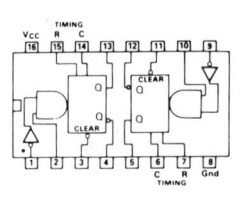

125

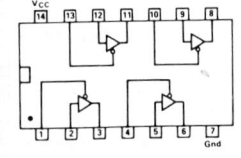

126

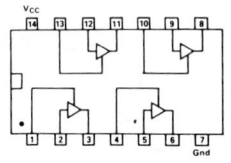

42 74 series

128

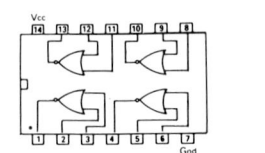

132

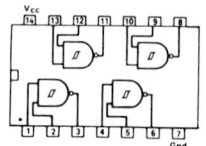

133

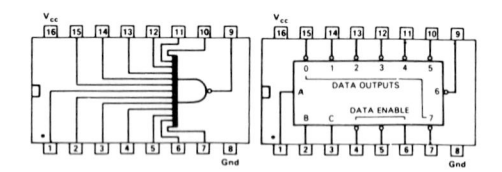

137

138

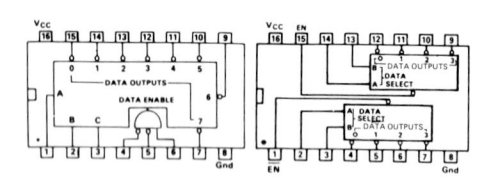

139

148

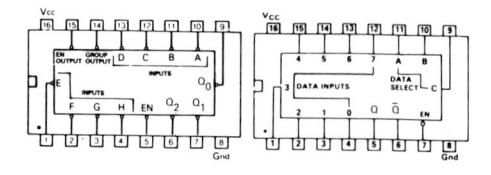

151

153

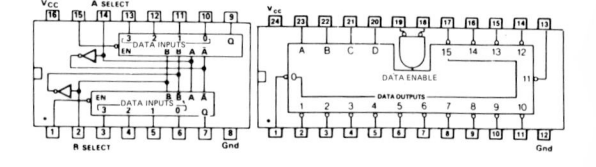

154

74 series

155

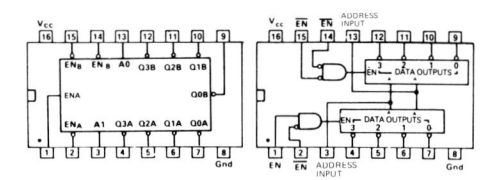

```
        Vcc
    ┌16─15─14─13─12─11─10──9┐
    │  ENB ENB A0 Q3B Q2B Q1B │
    │  ENA                 Q0B│
    │  ENA A1 Q3A Q2A Q1A Q0A │
    └─1──2──3──4──5──6──7──8─┘
                          Gnd
```

156

```
        Vcc  EN  EN   ADDRESS INPUT
    ┌16─15─14─13─12─11─10──9┐
    │         EN─ DATA OUTPUTS │
    │         EN─ DATA OUTPUTS │
    └─1──2──3──4──5──6──7──8─┘
       EN  EN  ADDRESS INPUT  Gnd
```

157

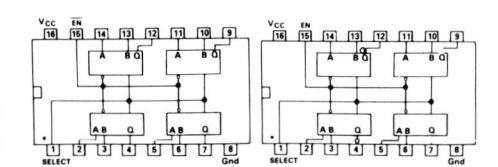

```
        Vcc  EN
    ┌16─15─14─13─12─11─10──9┐
    │       A  B  Q     A  B  Q │
    │       A  B  Q     A  B  Q │
    └─1──2──3──4──5──6──7──8─┘
      SELECT                  Gnd
```

158

```
        Vcc  EN
    ┌16─15─14─13─12─11─10──9┐
    │       A  B  Q     A  B  Q │
    │       A  B  Q     A  B  Q │
    └─1──2──3──4──5──6──7──8─┘
      SELECT                  Gnd
```

164

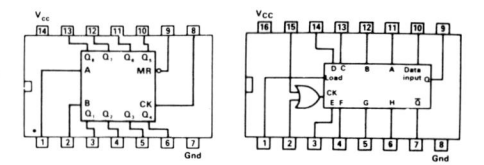

```
        Vcc
    ┌14─13─12─11─10──9──8┐
    │   Q4 Q5 Q6 Q7 MR   │
    │   A                │
    │   B  Q0 Q1 Q2 Q3 CK│
    └─1──2──3──4──5──6──7┘
                      Gnd
```

165

```
        Vcc
    ┌16─15─14─13─12─11─10──9┐
    │  D  C   B  A  Data     │
    │       Load       input │
    │       CK               │
    │  E  F  G  H          Q │
    └─1──2──3──4──5──6──7──8─┘
                           Gnd
```

174

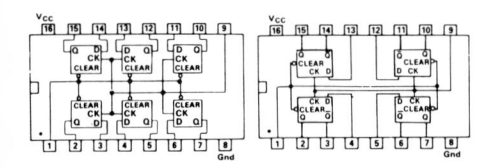

```
        Vcc
    ┌16─15─14─13─12─11─10──9┐
    │  D Q   D Q    D Q     │
    │  CLEAR CK  CK CLEAR   │
    │  CK    CLEAR CK       │
    │  CLEAR CLEAR CLEAR    │
    │  CK    CK    CK       │
    │  D Q   D Q    D Q     │
    └─1──2──3──4──5──6──7──8─┘
                           Gnd
```

175

```
        Vcc
    ┌16─15─14─13─12─11─10──9┐
    │  Q Q   Q Q            │
    │  CLEAR  D CK  D  CLEAR│
    │  CK    CLEAR CK       │
    │  CLEAR CK    CLEAR    │
    │  CK D  CLEAR  D CK    │
    │  Q Q   Q Q            │
    └─1──2──3──4──5──6──7──8─┘
                           Gnd
```

180

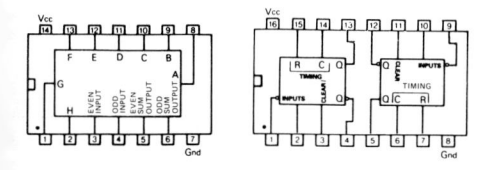

```
        Vcc
    ┌14─13─12─11─10──9──8┐
    │  F  E  D  C  B  A  │
    │  G                 │
    │  H EVEN ODD EVEN ODD
    │    INPUT INPUT SUM SUM
    │                OUT OUT
    └─1──2──3──4──5──6──7┘
                       Gnd
```

221

```
        Vcc
    ┌16─15─14─13─12─11─10──9┐
    │     R  C  Q       Q   INPUTS│
    │           TIMING        │
    │  INPUTS CLEAR     TIMING│
    │     Q  C  R           │
    └─1──2──3──4──5──6──7──8─┘
                           Gnd
```

44 74 series

240

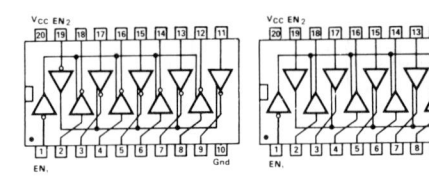

241

242

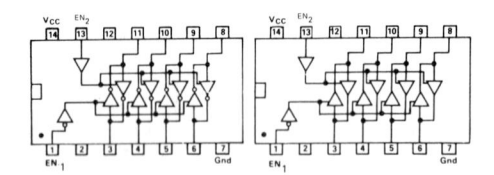

243

244

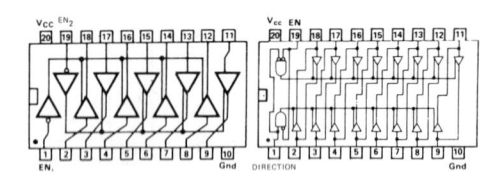

245

251

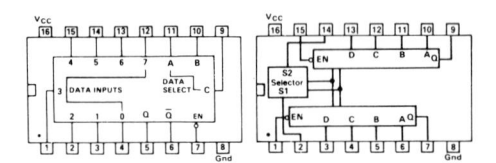

253

256

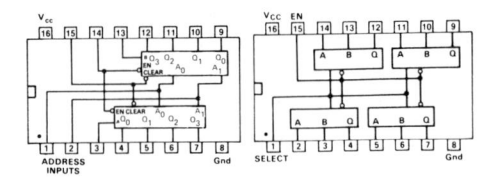

257

74 series 45

258

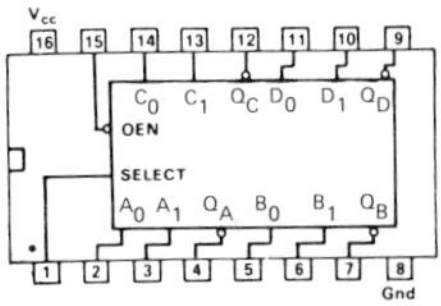

```
Vcc
[16] [15] [14] [13] [12] [11] [10] [9]
     C₀  C₁  O_C D₀  D₁  Q_D
     OEN
     SELECT
     A₀  A₁  Q_A B₀  B₁  Q_B
[1]  [2]  [3]  [4]  [5]  [6]  [7]  [8]
                                    Gnd
```

259

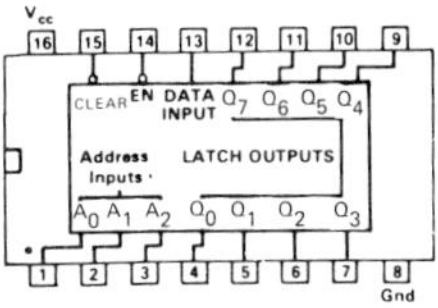

```
Vcc
[16] [15] [14] [13] [12] [11] [10] [9]
        EN  DATA
CLEAR   INPUT  Q₇ Q₆ Q₅ Q₄
Address Inputs   LATCH OUTPUTS
A₀ A₁ A₂  Q₀ Q₁ Q₂ Q₃
[1]  [2]  [3]  [4]  [5]  [6]  [7]  [8]
                                    Gnd
```

266

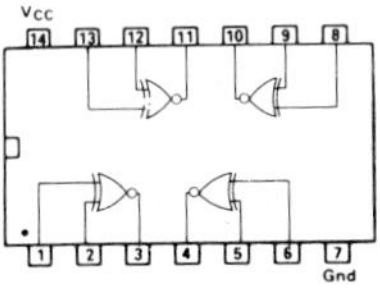

```
Vcc
[14] [13] [12] [11] [10] [9] [8]
[1]  [2]  [3]  [4]  [5]  [6]  [7]
                                Gnd
```

273

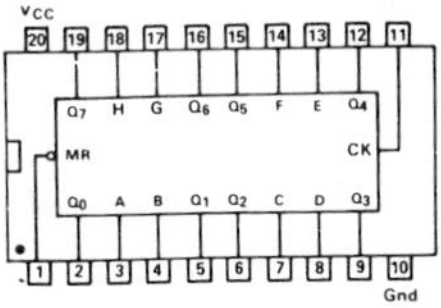

```
Vcc
[20] [19] [18] [17] [16] [15] [14] [13] [12] [11]
     Q₇  H   G   Q₅  Q₆  F   E   Q₄
MR                                      CK
     Q₀  A   B   Q₁  Q₂  C   D   Q₃
[1]  [2]  [3]  [4]  [5]  [6]  [7]  [8]  [9]  [10]
                                              Gnd
```

280

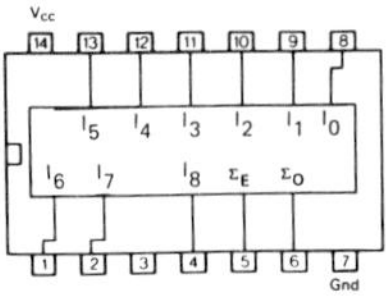

```
Vcc
[14] [13] [12] [11] [10] [9] [8]
  I₅  I₄  I₃  I₂  I₁  I₀
  I₆  I₇  I₈  Σ_E Σ_O
[1]  [2]  [3]  [4]  [5]  [6]  [7]
                                Gnd
```

298

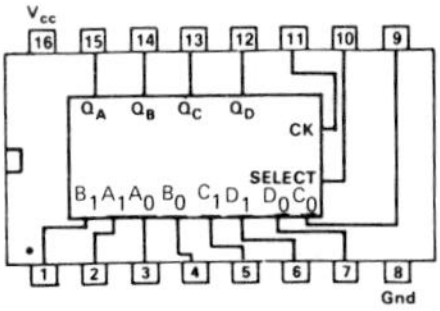

```
Vcc
[16] [15] [14] [13] [12] [11] [10] [9]
     Q_A Q_B Q_C Q_D   CK
                       SELECT
B₁ A₁ A₀ B₀ C₁ D₁ D₀ C₀
[1]  [2]  [3]  [4]  [5]  [6]  [7]  [8]
                                    Gnd
```

299

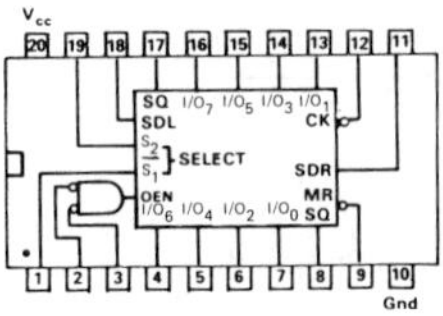

```
Vcc
[20] [19] [18] [17] [16] [15] [14] [13] [12] [11]
     SQ  I/O₇ I/O₅ I/O₃ I/O₁ CK
     SDL
     S₂
     S₁  SELECT
            SDR
     OEN                        MR
     I/O₆ I/O₄ I/O₂ I/O₀ SQ
[1]  [2]  [3]  [4]  [5]  [6]  [7]  [8]  [9]  [10]
                                              Gnd
```

323

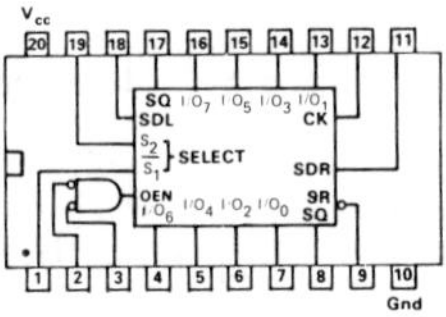

```
Vcc
[20] [19] [18] [17] [16] [15] [14] [13] [12] [11]
     SQ  I/O₇ I/O₅ I/O₃ I/O₁ CK
     SDL
     S₂  SELECT
     S₁          SDR
     OEN
     I/O₆ I/O₄ I/O₂ I/O₀ SQ   SR
[1]  [2]  [3]  [4]  [5]  [6]  [7]  [8]  [9]  [10]
                                              Gnd
```

352

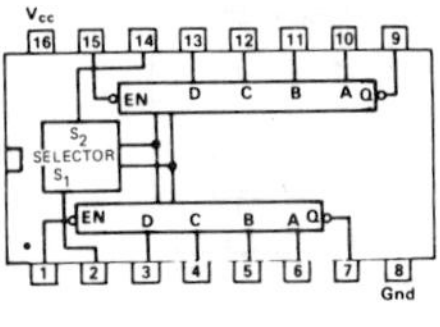

```
Vcc
[16] [15] [14] [13] [12] [11] [10] [9]
     EN  D   C   B   A   Q
     S₂
     SELECTOR
     S₁
     EN  D   C   B   A   Q
[1]  [2]  [3]  [4]  [5]  [6]  [7]  [8]
                                    Gnd
```

353

```
Vcc
[16] [15] [14] [13] [12] [11] [10] [9]
     EN  D   C   B   A   Q
     S₂
     SELECTOR
     S₁
     EN  D   C   B   A   Q
[1]  [2]  [3]  [4]  [5]  [6]  [7]  [8]
                                    Gnd
```

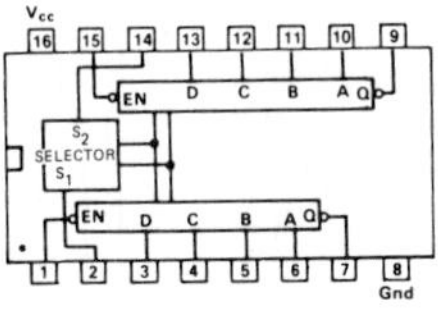

46 74 series

354

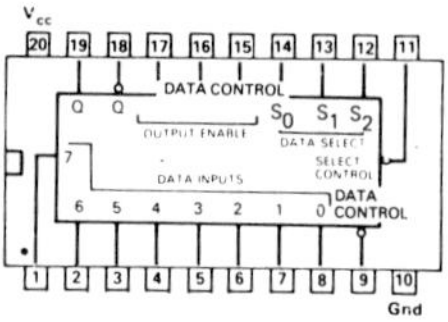

```
Vcc  20 19 18 17 16 15 14 13 12 11
         Q  Q    DATA CONTROL  S0 S1 S2
              OUTPUT ENABLE    DATA SELECT
          7                     SELECT CONTROL
              DATA INPUTS       DATA
          6  5  4  3  2  1   0  CONTROL
      1  2  3  4  5  6  7  8  9 10
                              Gnd
```

356

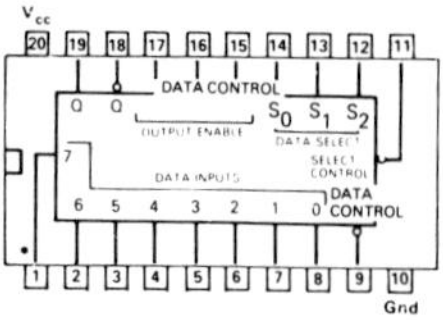

```
Vcc  20 19 18 17 16 15 14 13 12 11
         Q  Q    DATA CONTROL  S0 S1 S2
              OUTPUT ENABLE    DATA SELECT
          7                     SELECT CONTROL
              DATA INPUTS       DATA
          6  5  4  3  2  1   0  CONTROL
      1  2  3  4  5  6  7  8  9 10
                              Gnd
```

365

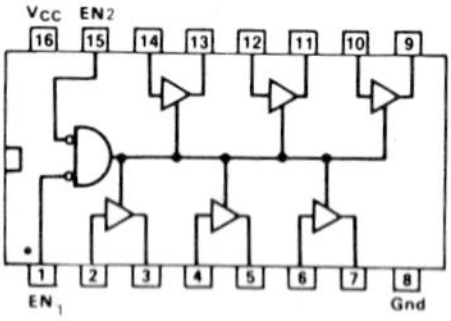

```
Vcc EN2
 16  15  14  13  12  11  10  9
  1   2   3   4   5   6   7  8
 EN1                         Gnd
```

366

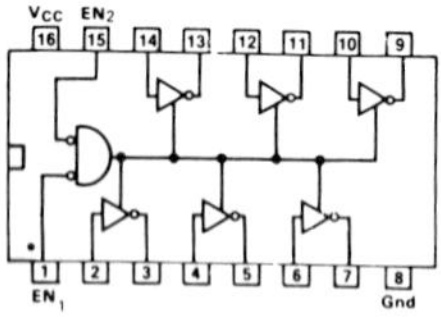

```
Vcc EN2
 16  15  14  13  12  11  10  9
  1   2   3   4   5   6   7  8
 EN1                         Gnd
```

367

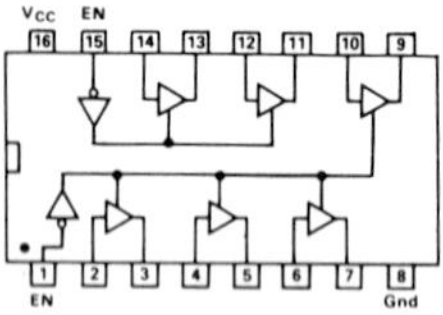

```
Vcc EN
 16  15  14  13  12  11  10  9
  1   2   3   4   5   6   7  8
 EN                          Gnd
```

368

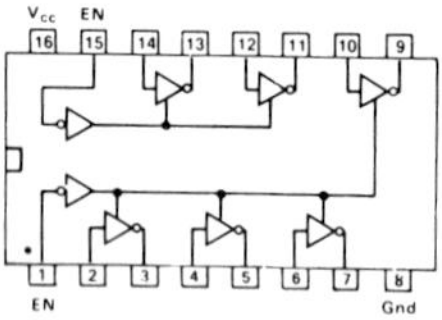

```
Vcc EN
 16  15  14  13  12  11  10  9
  1   2   3   4   5   6   7  8
 EN                          Gnd
```

373

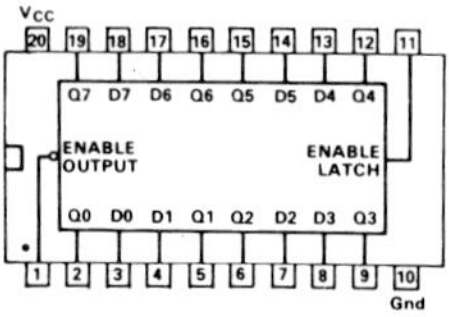

```
Vcc 20 19 18 17 16 15 14 13 12 11
        Q7 D7 D6 Q6 Q5 D5 D4 Q4
    ENABLE              ENABLE
    OUTPUT               LATCH
        Q0 D0 D1 Q1 Q2 D2 D3 Q3
     1  2  3  4  5  6  7  8  9 10
                              Gnd
```

374

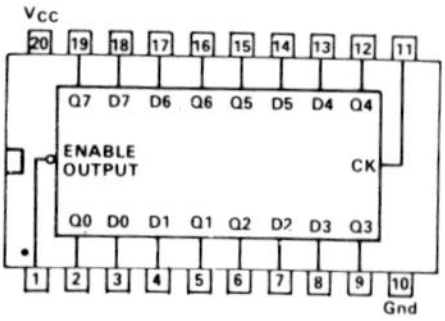

```
Vcc 20 19 18 17 16 15 14 13 12 11
        Q7 D7 D6 Q6 Q5 D5 D4 Q4
    ENABLE                     CK
    OUTPUT
        Q0 D0 D1 Q1 Q2 D2 D3 Q3
     1  2  3  4  5  6  7  8  9 10
                              Gnd
```

377

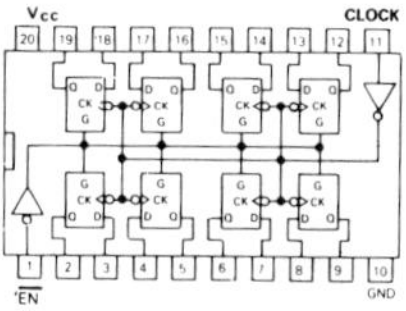

```
Vcc                          CLOCK
 20 19 18 17 16 15 14 13 12 11
  1  2  3  4  5  6  7  8  9 10
 EN                          GND
```

378

```
Vcc
 16  15  14  13  12  11  10  9
        Q5  F   E   Q4  D   Q3
    EN                       CK
        Q0  A   B   Q1  C   Q2
  1   2   3   4   5   6   7  8
                             Gnd
```

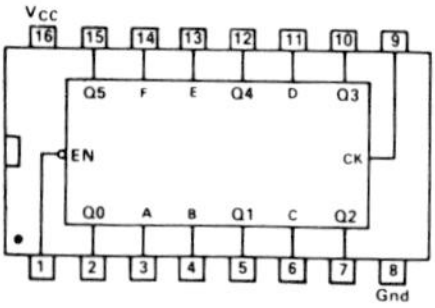

74 series 47

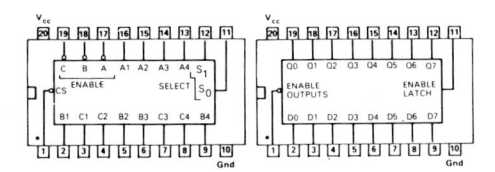

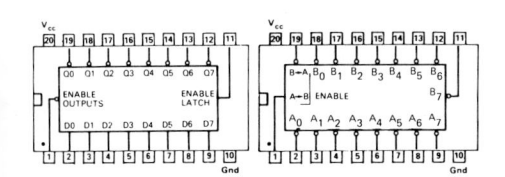

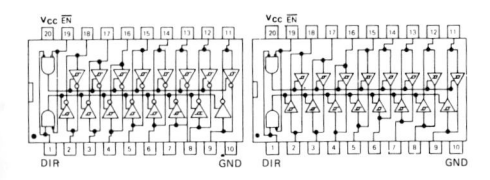

442

573

580

620

640

643

673

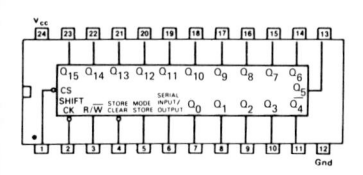

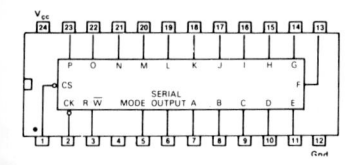

674

4000 Series

- 4001 Quad 2-input NOR
- 4002 Dual 4-input NOR
- 4006 18-bit shift register
- 4007 Dual CMOS transistor pair/inverter
- 4008 4-bit full adder
- 4009 Hex inverter/buffer
- 4010 Hex buffer
- 4011 Quad 2-input NAND
- 4012 Dual 4-input NAND
- 4013 Dual D-type bistable
- 4014 8-bit shift register
- 4015 Dual 4-bit shift register
- 4016 Quad analogue switch
- 4017 Decade counter
- 4018 Divide-by-N counter
- 4019 Quad 2-input AND/OR
- 4020 14-bit binary counter
- 4021 8-bit shift register
- 4022 Octal counter
- 4023 Triple 3-input NAND
- 4024 Seven-stage ripple counter
- 4025 Triple 3-input NOR
- 4026 7-segment display driver
- 4027 Dual J-K bistable
- 4028 BCD to decimal/binary to octal decoder
- 4029 Presettable binary/BCD up/down counter
- 4030 Quad 2-input exclusive-OR
- 4031 64-bit shift register
- 4032 Triple serial adder
- 4033 7-segment display driver
- 4034 8-bit bi-directional shift register
- 4035 4-bit PIPO shift register
- 4036 32-bit RAM
- 4037 Triple 3-input AND/OR
- 4038 Triple serial adder
- 4039 32-bit RAM
- 4040 12-bit binary counter
- 4041 Quad inverter/buffer
- 4042 Quad D-type latch
- 4043 Quad tri-state R-S latch
- 4044 Quad tri-state R-S latch
- 4045 21-bit binary counter
- 4046 Phase-locked loop
- 4047 Monostable/astable
- 4048 8-input multifunction gate
- 4049 Hex inverter/buffer
- 4050 Hex buffer
- 4051 Single 8-input analogue multiplexer
- 4052 Dual 4-input analogue multiplexer
- 4053 Triple 2-input analogue multiplexer
- 4054 BCD 7-segment display decoder/LCD driver
- 4055 BCD 7-segment display decoder/LCD driver
- 4056 BCD 7-segment display decoder/LCD driver
- 4057 Arithmetic logic unit
- 4059 Divide-by-N counter

4000 series

- 4060 14-bit binary counter
- 4061 256-bit RAM
- 4062 200-bit shift register
- 4063 4-bit magnitude comparator
- 4066 Quad analogue switch
- 4067 1 to 16-line multiplexer/demultiplexer
- 4068 Single 8-input NAND
- 4069 Hex inverter
- 4070 Quad exclusive-OR
- 4071 Quad 2-input OR
- 4072 Dual 4-input OR
- 4073 Triple 3-input AND
- 4075 Triple 3-input OR
- 4076 Quad D-type register
- 4077 Quad 2-input exclusive-NOR
- 4078 Single 8-input NOR
- 4081 Quad 2-input AND
- 4082 Dual 4-input AND
- 4085 Dual 2-input AND/OR/invert
- 4086 Dual 2-input AND/OR/invert
- 4089 Binary rate multiplier
- 4093 Quad 2-input NAND
- 4094 8-stage tri-state register
- 4095 Single J-K bistable
- 4096 Single J-K bistable
- 4097 Dual 8-channel multiplexer/demultiplexer
- 4098 Dual retriggerable monostable
- 4099 8-bit addressable latch
- 4104 Quad level translator
- 4160 4-bit programmable decade counter
- 4161 4-bit programmable binary counter
- 4162 4-bit programmable decade counter
- 4163 4-bit programmable binary counter
- 4174 Hex D-type bistable
- 4175 Quad D-type bistable
- 4194 4-bit bi-directional shift register

4000 Series pin connections

4001

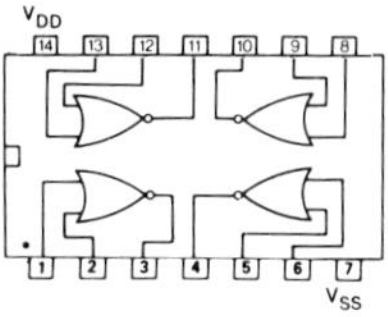

4002

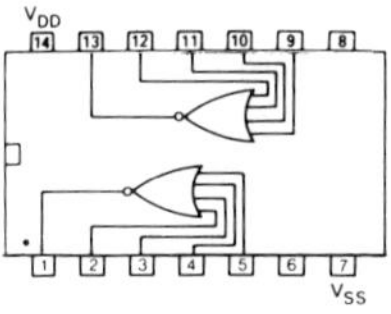

4011

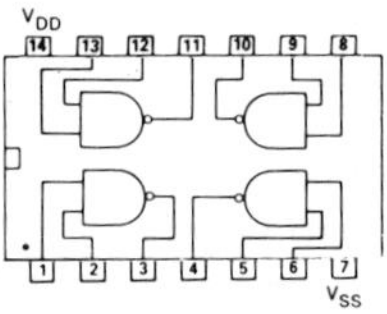

4012

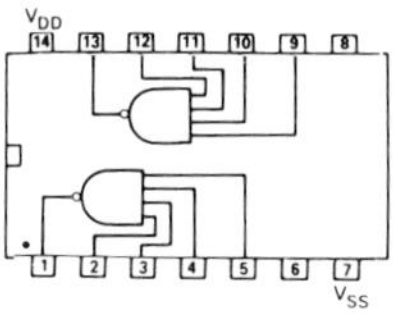

50 4000 series

4013

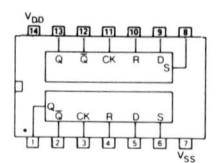

4023

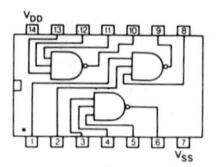

4042

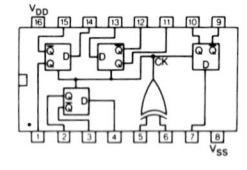

4049

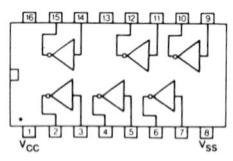

4050

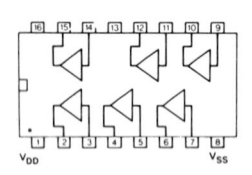

4068

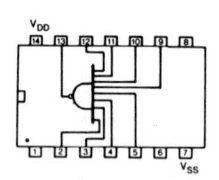

4069

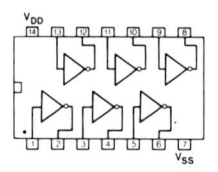

4070

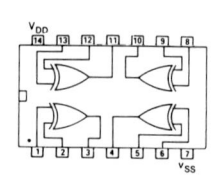

4071

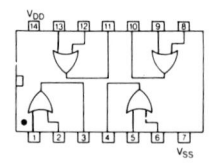

4072

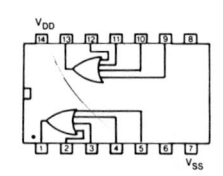

4073

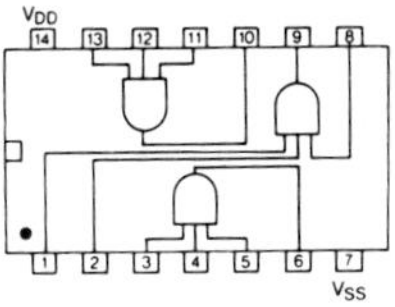

4075

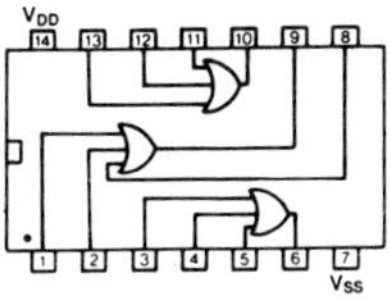

4076

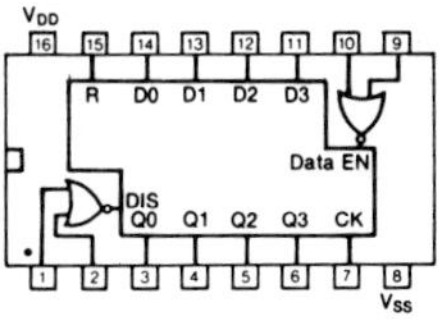

4077

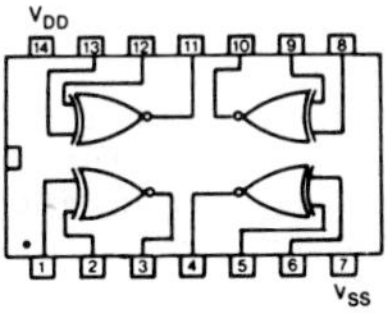

4078

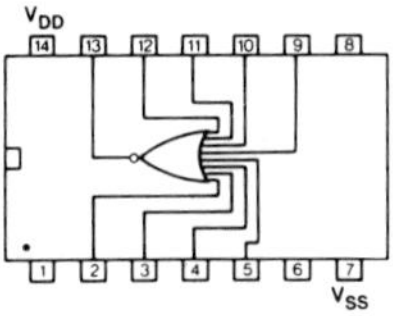

4081

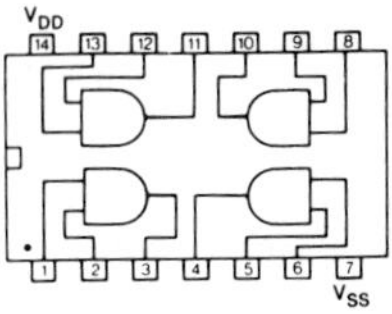

4093

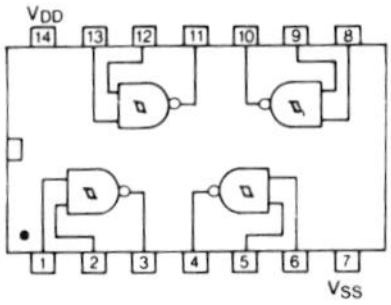

4500 Series

- 4501 Dual 4-input NAND/single 2-input OR/NOR
- 4502 Hex str. inverter/buffer
- 4503 Hex tri-state buffer
- 4504 Hex TTL-CMOS level shifter
- 4505 64-bit RAM
- 4506 Dual AND/OR/invert
- 4507 Quad exclusive-OR

4500 series

4508	Dual 4-bit tri-state latch
4510	BCD up/down counter
4511	BCD to 7-segment latch/decoder/driver
4512	8-channel data selector
4513	BCD to 7-segment latch/decoder/driver
4514	4-bit latched input 1 to 16-line decoder
4515	4-bit latched input 1 to 16-line decoder
4516	Binary up/down counter
4517	Dual 64-bit shift register
4518	Dual BCD up-counter
4519	Quad 2-input multiplexer
4520	Dual 4-bit binary counter
4521	24-stage frequency divider
4522	BCD programmable divide-by-N
4524	256 × 4-bit ROM
4526	Binary programmable divide-by-N
4527	BCD rate multiplier
4528	Dual resettable monostable
4529	Dual 4-channel tri-state analogue data selector
4530	Dual 5-input majority gate
4531	12-bit parity tree
4532	8-bit priority encoder
4534	5-decade counter
4536	Programmable timer
4537	256 × 1-bit RAM
4538	Dual retriggerable/resettable monostable
4539	Dual 4-channel data selector
4541	Programmable timer
4543	BCD to 7-segment latch/decoder/driver
4544	BCD to 7-segment latch/decoder/driver
4547	BCD to 7-segment latch/decoder/driver
4549	Successive approximation register
4551	Quad 2-input analogue multiplexer
4552	64 × 4-bit RAM
4553	3-digit BCD counter
4554	2-bit binary multiplier
4555	Dual 2 to 4-line decoder
4556	Dual 2 of 4-line decoder
4557	Variable length shift register
4558	BCD to 7-segment latch decoder
4559	Successive approximation register
4560	4-bit BCD adder
4561	9's complementer
4562	128-bit shift register
4566	Time base generator
4568	Programmable counter/phase comparator
4569	Dual programmable BCD counter
4572	2-input AND/2-input NOR/quad inverter
4573	Quad programmable operational amplifier
4574	Quad programmable comparator
4575	Dual operational amplifier/dual comparator
4580	4 × 4 multi-port register
4581	4-bit arithmetic logic unit
4582	Look-ahead carry generator
4583	Dual Sch. trigger
4584	Hex Sch. inverter
4585	4-bit magnitude comparator
4597	8-bit bus compatible latch
4598	8-bit bus compatible latch
4599	8-bit addressable latch

4500 Series pin connections

4502

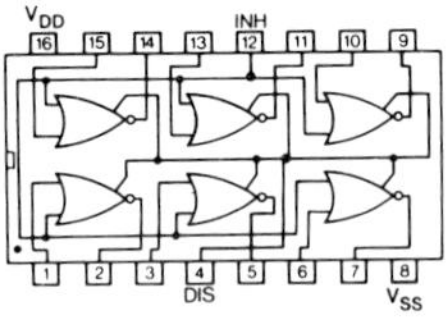

4508

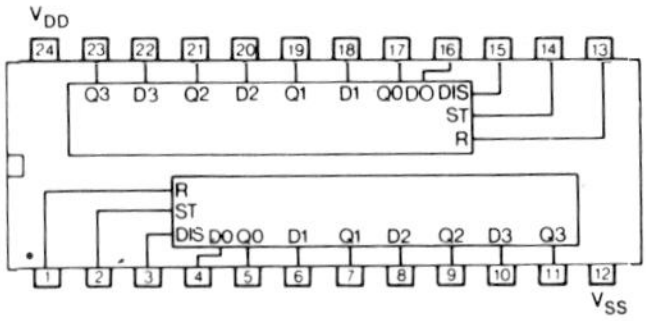

4512

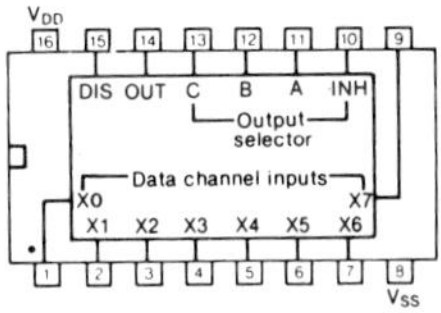

4514

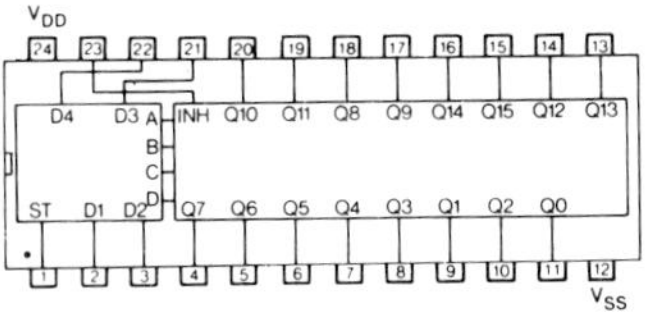

4515

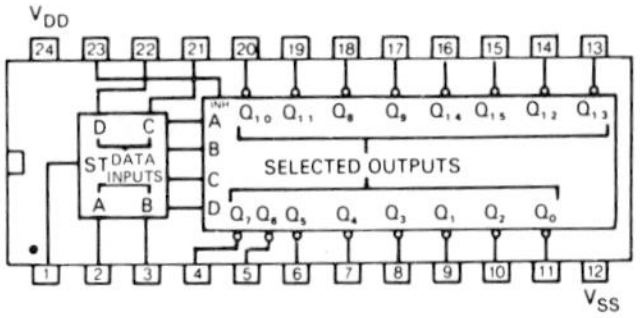

4597

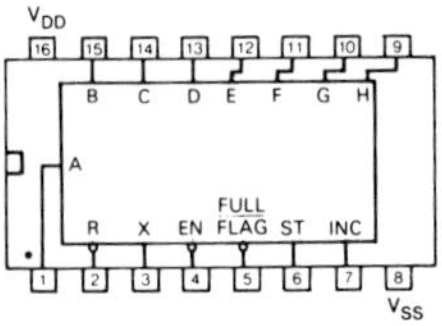

4598

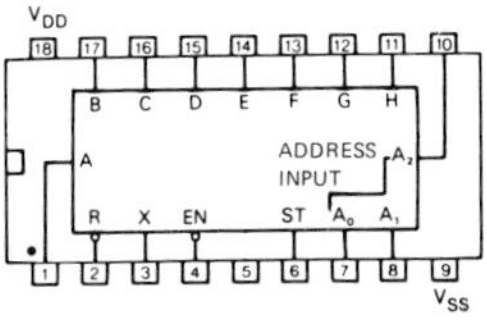

4599

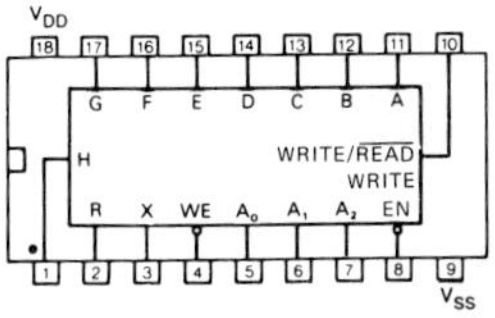

Electrical characteristics of typical logic gates

Family	Operating supply voltage (V)	Power consumption (per gate) (mW)	Average propagation delay (ns)	Fan-out (same family)	Maximum switching frequency (MHz)
7400	4.75–5.25	10	10	10	30
74C	3–15	*	90	10	2.5
74F	4.75–5.25	6	4	20	120
74H	4.75–5.25	20	6	10	50
74L	4.75–5.25	1	40	10	3
74S	4.75–5.25	20	5	10	100
74HC	3–6	*	12	10	25
74LS	4.75–5.25	2	10	20	30
74ALS	4.5–5.5	1	5	20	35
74HCT	4.5–5.5	*	15	20	25
4000A	3–15	*	150	10	1
4000B	3–18	*	200	10	2.5

*Depends on frequency: typically 20 μW at 10 kHz, 2 mW at 2 MHz.

Fan-in and fan-out of logic gates

The fan-in of a logic gate provides a measure of the loading effect presented by its inputs and is usually expressed as the number of standard loads that it represents.

The fan-out of a logic gate provides a measure of the number of standard logic gate inputs that may be connected without the logic levels becoming illegal.

Typical values of fan-in and fan-out, expressed in terms of a standard TTL load, are shown below:

Family	Fan-in Low state	Fan-in High State	Fan-out Low state	Fan-out High state
7400	1	1	10	20
74F	0.375	0.5	12.5	25
74HC	0.05	0.05	2.5	2.5
74LS	0.25	0.5	5	10
74ALS	0.06	0.5	5	10
74HCT	0.05	0.05	2.5	2.5
4000	0.025	0.025	0.5	0.5

Standard TTL load

A standard TTL load (unit load) may be defined as:
High state input current = 40 μA
Low state input current = −1.6 mA
Typical conditions are illustrated below:

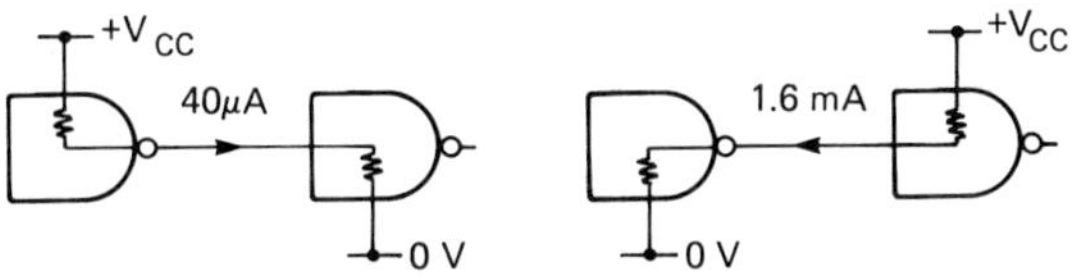

(a) High state input

(b) Low state input

TTL input and output current

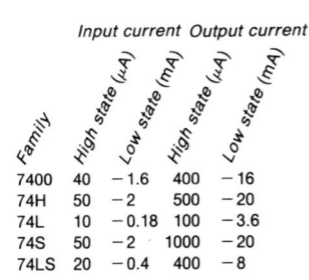

Family	Input current High state (μA)	Low state (mA)	Output current High state (μA)	Low state (mA)
7400	40	−1.6	400	−16
74H	50	−2	500	−20
74L	10	−0.18	100	−3.6
74S	50	−2	1000	−20
74LS	20	−0.4	400	−8

Interconnecting TTL families

Maximum number of inputs that may be connected

Family	7400	74H	74L	74S	74LS
74	10	8	40	8	20
74H	12	10	40	10	25
74L	2	1	10	1	5
74S	12	10	40	10	50
74LS	5	4	20	4	20

Logic levels and noise margins for CMOS and TTL

Logic levels are the range of voltages used to represent the logic states 0 and 1. With conventional positive logic these are as follows:

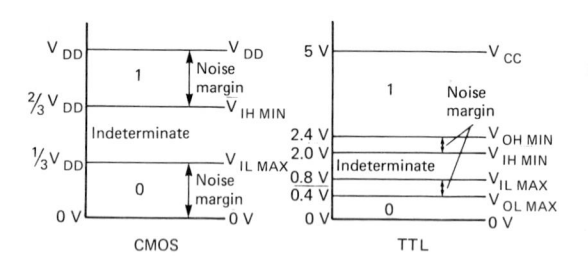

```
CMOS                                    TTL
V_DD ─────── V_DD              5 V ───────── V_CC
         ↑ Noise
   1     ↓ margin                    1      Noise
2/3 V_DD ──── V_IH MIN                       margin
                               2.4 V ─────── V_OH MIN
 Indeterminate                 2.0 V ─────── V_IH MIN
1/3 V_DD ──── V_IL MAX        Indeterminate
         ↑ Noise               0.8 V ─────── V_IL MAX
   0     ↓ margin              0.4 V ─────── V_OL MAX
  0 V ────── 0 V                 0 V    0    0 V
```

The noise margin is defined as the difference between:
(i) the minimum values of high state output and input voltage, $V_{O=MIN}$ and $V_{I=MIN}$.
(ii) the maximum values of low state output and input voltage, $V_{OL\ MAX}$ and $V_{IL\ MAX}$.

The noise margin for standard 7400-series TTL is usually 400 mV while that for CMOS is 0.33 V_{DD}.

Boolean algebra

Boolean operators
. = AND   + = OR   ⊕ = exclusive-OR   ¯ = NOT

Boolean identities
0.0 = 0   0 + 0 = 0
0.1 = 0   0 + 1 = 1
1.0 = 0   1 + 0 = 1
1.1 = 1   1 + 1 = 1

A.0 = 0   A + 0 = A
A.1 = A   A + 1 = 1
A.A = A   A + A = A
A.$\bar{A}$ = 0   A + $\bar{A}$ = 1

$\bar{\bar{A}}$ = A $\bar{\bar{\bar{A}}}$ = $\bar{A}$

A + A.B = A   A + $\bar{A}$.B = A + B
(A + B).(A + C) = A + B.C

A ⊕ B = A.$\bar{B}$ + $\bar{A}$.B

Associative law
A + (B + C) = (A + B) + C   and   A.(B.C) = (A.B).C
           = A + B + C   = A.B.C

Commutative law
$A+B = B+A$ and $A.B = B.A$

Distributive law
$A.(B+C) = A.B+A.C$ and $A+(B.C) = (A+B).(A+C)$

De Morgan's theorem

$$\overline{A.B} = \overline{A}+\overline{B} \qquad \overline{A+B} = \overline{A}.\overline{B}$$
$$\overline{A.B.C} = \overline{A}+\overline{B}+\overline{C} \qquad \overline{A+B+C} = \overline{A}.\overline{B}.\overline{C}$$

Karnaugh maps

Karnaugh maps are a useful graphical technique for simplifying complex logical functions involving between two and eight variables; beyond that it is better to employ computer simulation.

The Karnaugh map consists of a square or rectangular array of cells into which 0s and 1s may be placed to indicate false and true respectively. Alternative representations of a Karnaugh map for two variables are shown below:

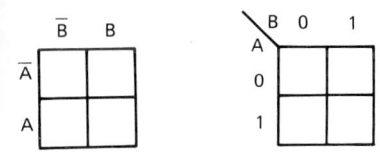

The relationship between a truth table and a Karnaugh map is illustrated in the following example, which plots the AND function:

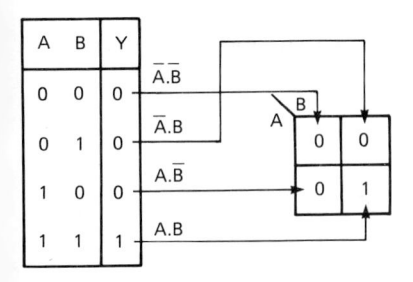

A	B	Y
0	0	0
0	1	0
1	0	0
1	1	1

Karnaugh maps for the remaining basic logic functions (NAND, OR and NOR) for two variables are shown below:

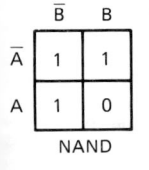

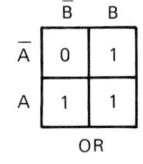

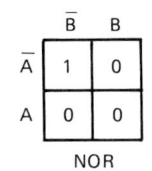

NAND   OR   NOR

58 Karnaugh maps

Adjacent cells within a Karnaugh map may be grouped together in rectangles of two, four, eight, etc. cells in order to effect a simplification.

Taking the NAND function, for example, the two groups of two adjacent cells in the map (below) correspond to $\overline{A}$ and $\overline{B}$. We thus conclude that:

$\overline{A.B} = \overline{A} + \overline{B}$ (De Morgan's theorem)

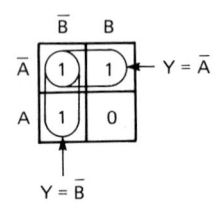

	$\overline{B}$	B
$\overline{A}$	1	1
A	1	0

$Y = \overline{A}$ (top row)
$Y = \overline{B}$ (left column)

The technique of grouping cells together is an extremely powerful one. On a Karnaugh map showing four variables the relationship between the number of cells grouped together and the number of variables is as follows:

No. of cells	No. of variables
1	4
2	3
4	2
8	1

The following example shows how the function

$Y = A.B.C.D + A.B.C.\overline{D} + A.B.\overline{C}.D + A.B.\overline{C}.\overline{D} + A.\overline{B}.C.D + A.\overline{B}.C.\overline{D}$

reduces to

$Y = A.B + A.C$

	$\overline{C}.\overline{D}$	$\overline{C}.D$	$C.D$	$C.\overline{D}$
$\overline{A}.\overline{B}$	0	0	0	0
$\overline{A}.B$	0	0	0	0
$A.B$	1	1	1	1
$A.\overline{B}$	0	0	1	1

It is also important to note that the map is a continuous surface which links edge to edge. This allows cells at opposite extremes of a row or column to be linked together. The four corner cells may likewise be grouped together (provided they all contain 1!).

Some possible cell groupings are shown below:

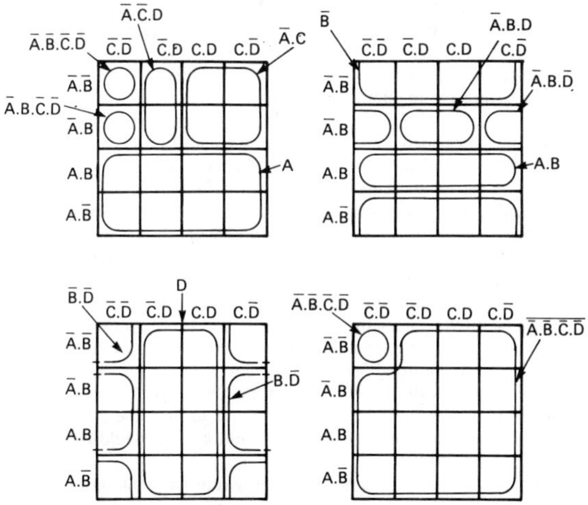

Power supplies

Most TTL and CMOS logic systems are designed to operate from a single supply rail of nominally +5 V. This voltage should be regulated to within ±5% (i.e. it should not be allowed to fall outside the range 4.75 V to 5.25 V) and the impedance of the supply must be very low (typically 0.1 ohm or less) over a wide range of frequencies (up to 35 MHz for standard TTL and CMOS and up to 150 MHz for 'fast' and Schottky TTL).

The +5 V supply may be conveniently derived from a monolithic three-terminal regulator as shown below. Regulators of this type normally require an adequate heat sink (of around 4°C/W) and should be fitted with high-frequency decoupling capacitors (of typically 220 nF). These should be mounted as close to the regulator's terminals as possible.

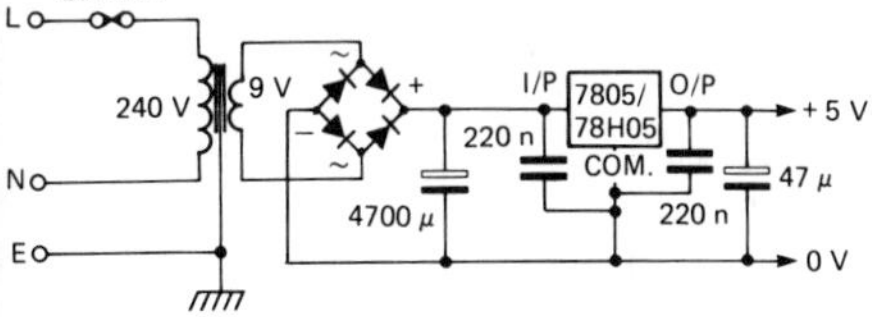

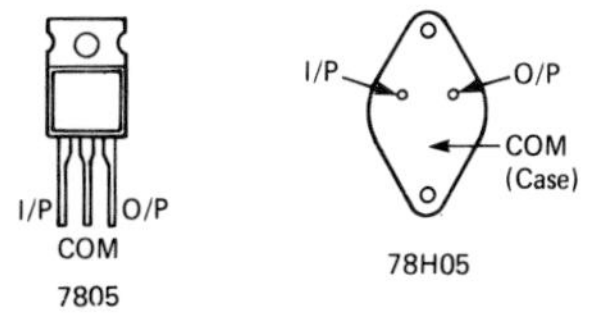

Under no circumstances should a TTL supply rail be allowed to exceed +7 V as this is likely to cause permanent damage to the integrated circuits. More elaborate power supplies may incorporate 'crowbar' protection in order to combat the effects of short-circuit failure within the regulator.

Care must also be exercised with supply distribution to the individual integrated circuits. Main +5 V and 0 V rails should use PCB tracks of at least 5 mm width and generous 0 V 'land' areas should be provided. (These can also be useful as a means of heat conduction from integrated circuits soldered directly to the PCB.)

Supply connectors and interconnecting leads should be adequately rated and all supply connections should be kept as short and direct as possible.

High-frequency decoupling capacitors (e.g. disk ceramic types) should be fitted as near to the individual i.c. supply pins as possible. At least one such capacitor (of between 4.7 nF and 100 nF) should be fitted for every three to four i.c. devices.

Low-frequency decoupling capacitors (electrolytic) should be fitted to main supply rails at regular points around the PCB. At least one capacitor (of between 4.7 μF and 47 μF) should be fitted for every eight to ten i.c. devices.

Transient 'spikes' occurring on supply rails can manifest themselves in various ways, including spurious data errors and system crashes. Spikes can readily be detected using an oscilloscope (connected at various points to the +5 V rail) and dealt with by reinforcing the supply rail decoupling at strategic points. In severe cases ferrite bead inductors may be necessary in order to provide more effective decoupling.

Interfacing logic families

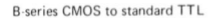

B-series CMOS to standard TTL

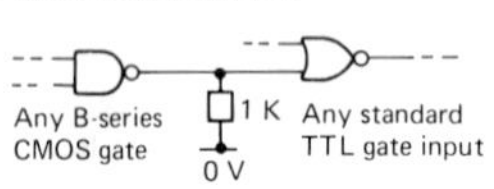

Any B-series CMOS gate — 1 K to 0 V — Any standard TTL gate input

CMOS to LS-TTL

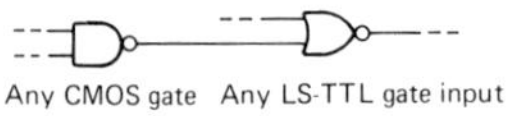

Any CMOS gate — Any LS-TTL gate input

TTL to CMOS

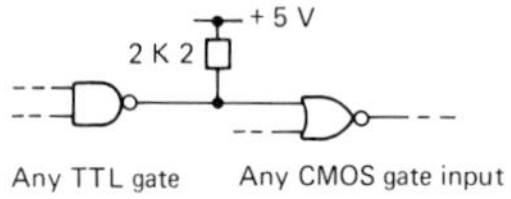

Any TTL gate — 2 K2 to +5 V — Any CMOS gate input

CMOS buffer to TTL

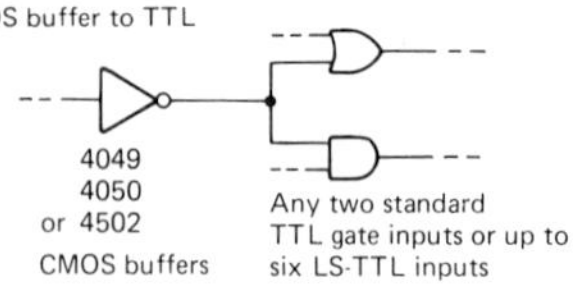

4049
4050
or 4502
CMOS buffers — Any two standard TTL gate inputs or up to six LS-TTL inputs

Microcomputer architecture

The essential constituents of any microcomputer system are:
(a) a central processing unit (CPU) which generally takes the form of a single VLSI device, the microprocessor;
(b) a memory which invariably comprises both read/write and read only devices (RAM and ROM respectively); and
(c) interface devices to facilitate input and output (I/O) for peripheral devices such as keyboards, disk drives, monitors and printers.

The individual elements of a microcomputer system are interconnected by means of a multiple connecting system known as a bus. In most systems there are three distinct buses: the address bus, the data bus, and the control bus.

The address bus is used to specify the memory locations (addresses) involved in data transfer while the data itself is transferred between devices using the data bus. The data bus, therefore, must be bidirectional — allowing data to be 'read' into and to be 'written' from the CPU.

The control bus comprises various lines used to distribute timing and control signals throughout the system. Important among these are: signals concerned with the direction of data transfer (to or from the CPU); signals which indicate that data is to be transferred to I/O rather than memory; and requests from external devices that require the attention of the CPU. The response to such 'interrupts' can be programmed in various ways, and a system of prioritization may often be desirable.

A system clock generator is responsible for providing an accurate and highly stable timing signal. This generator often forms part of the microprocessor itself.

The number of lines contained in the address and data buses depend upon the particular microprocessor employed. Most of today's microprocessors are capable of performing operations on binary numbers consisting of either 8 or 16 bits. They are thus known as 8-bit and 16-bit microprocessors respectively.

In a microcomputer based on an 8-bit microprocessor, the data bus has 8 separate lines. Similarly, in a 16-bit system the data bus will have 16 separate lines. Address buses for 8-bit systems invariably comprise 16 lines whereas those for 16-bit systems may consist of as many as 24 lines.

A further complication exists in the case of a number of microprocessors which, in order to minimize the CPU pin count (so that a 40-pin rather than a 64-pin package may be utilized), employ multiplexed data and address buses. Certain CPU pins are then used to convey both address and data information, the CPU outputting a signal which is used to latch the multiplexed information onto the respective bus.

Since a bus may be connected to many devices, the use of bus drivers/buffers is often desirable in order to reduce the loading on the CPU. Bus drivers/buffers are usually packaged in groups of eight (for obvious reasons) and may be unidirectional (e.g. for use with an address bus) or bidirectional (e.g. for use with a data bus). In the latter case devices are usually referred to as 'bus transceivers'.

The largest binary number that can appear on an 8-bit bus is 11111111 (or $2^8 - 1$) while that for a 16-bit bus is 1111111111111111 (or $2^{16} - 1$). Each address corresponds to a unique binary code, hence the linear addressable range (i.e. the total number of memory locations available without 'paging') will be dependent upon the number of address lines provided within the system. (The maximum number of individual memory locations that can exist in a system having n address lines is 2^n.)

The relationship between the number of address lines and the linear address range for three popular microprocessors is shown below:

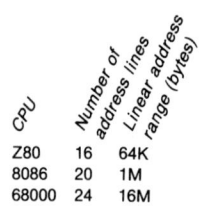

CPU	Number of address lines	Linear address range (bytes)
Z80	16	64K
8086	20	1M
68000	24	16M

Signals on all lines, whether they be address, data, or control, can exist in only one of two states: logic 0 (low) or logic 1 (high). As far as individual devices sharing the data bus are concerned, a third 'high impedance' state exists whenever a device is in its deselected or disabled state. This allows the CPU to communicate with other devices without the risk of a bus conflict. Bus transceivers can usually also be placed in a tri-state condition, thus permitting partial access to the bus for a second processor or other 'intelligent' device.

The address range corresponding to a particular device (e.g. ROM) is decoded from the address bus and used to generate an appropriate 'enable' signal. A TTL decoder (or demultiplexer) is often used in such an application.

Although the CPU is the heart of any microcomputer system, it may not be the only 'intelligent' device present. A second processor, for example, may be fitted, in order to perform numeric data processing (NDP), or a dedicated microprocessor may be incorporated in an 'intelligent keyboard'.

System VLSI devices other than CPU and memory (ROM and RAM) may include memory controllers, counter/timers, and serial/parallel I/O chips. Such support devices not only serve to simplify the task of the CPU but also help to minimize the overall chip count.

Simplified model of a microcomputer system

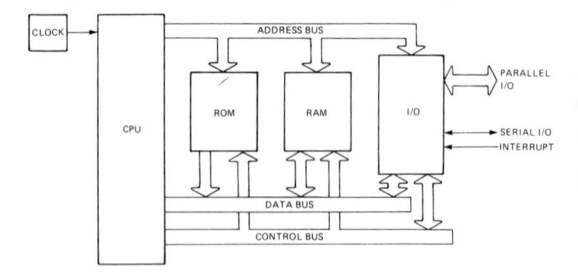

Typical 16-bit microcomputer system

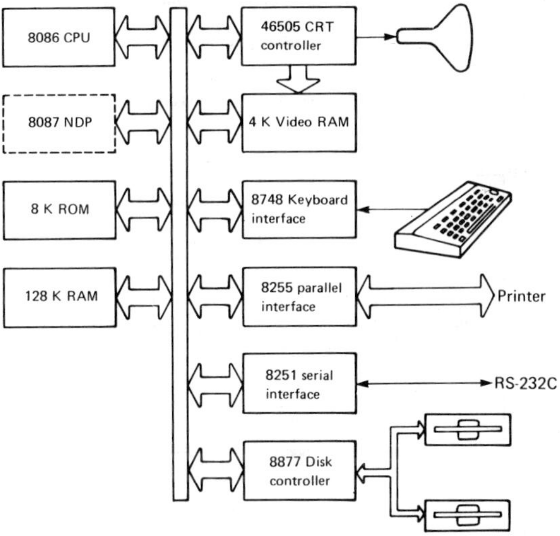

Memory maps

Portions of the address space comprising a microcomputer memory can be used in various ways. Some areas will be devoted to data and program storage, and must therefore have the capacity to be written to and read from (read/write memory), whereas other areas will be used for more permanent storage of the operating system, which will normally only permit reading (read only memory). In addition, on many home computers BASIC or other high level language interpreters are provided and these are invariably also contained in read only memory.

It is often convenient to think of a microcomputer memory as being divided into several contiguous blocks of appropriate size (e.g. 8K, 16K or 32K bytes). The resulting blocks can then be allocated various functions depending upon the particular system. A simple personal computer may, for example, have a total memory of 64K bytes, in which the operating system may exist in a block of 8K bytes, a BASIC interpreter may require a block of 16K bytes, and the remaining 40K bytes of memory may be filled by RAM. If 8K bytes of RAM are devoted to a bit-mapped screen, it should be apparent that a total of 32K bytes of RAM remains available for systems use, user programs and data.

In general, it is not necessary to draw memory maps to scale, nor is it necessary to show addresses in both decimal and hexadecimal format. Fortunately, most manufacturers provide memory maps for their systems and these can be a valuable aid to

64 Memory maps

the computer engineer. It should, however, be noted that there are significant differences between memory maps for machines intended primarily for games/home use and those intended for more serious applications. Similarly, systems using different microprocessors often have quite different memory maps. A Z80-based microcomputer, for example, will normally have ROM at the bottom of memory while a 6502-based machine will have ROM at the top of memory.

Typical memory map for an 8-bit business microcomputer running CP/M

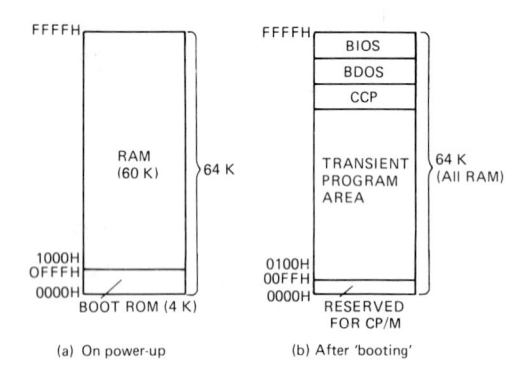

(a) On power-up

```
FFFFH ┌──────────────┐
      │              │
      │              │
      │   RAM        │
      │   (60 K)     │ } 64 K
      │              │
      │              │
1000H │              │
0FFFH ├──────────────┤
0000H │ BOOT ROM (4 K)│
      └──────────────┘
```

(b) After 'booting'

```
FFFFH ┌──────────────┐
      │    BIOS      │
      ├──────────────┤
      │    BDOS      │
      ├──────────────┤
      │    CCP       │
      ├──────────────┤
      │              │
      │  TRANSIENT   │ } 64 K
      │  PROGRAM     │  (All RAM)
      │  AREA        │
      │              │
0100H │              │
00FFH ├──────────────┤
0000H │ RESERVED     │
      │ FOR CP/M     │
      └──────────────┘
```

Typical memory map for a 16-bit personal computer running MS-DOS

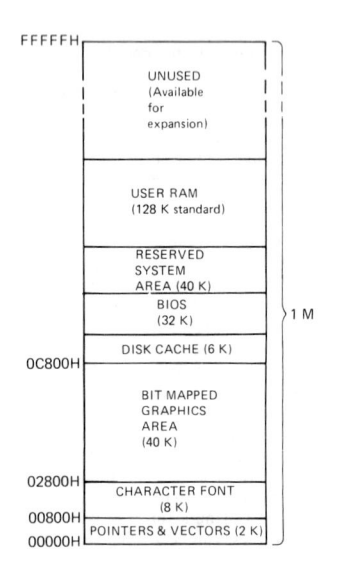

```
FFFFFH ┌──────────────────┐
       │    UNUSED        │
       │   (Available     │
       │      for         │
       │   expansion)     │
       ├──────────────────┤
       │   USER RAM       │
       │  (128 K standard)│
       ├──────────────────┤
       │   RESERVED       │
       │   SYSTEM         │
       │   AREA (40 K)    │
       ├──────────────────┤
       │    BIOS          │
       │    (32 K)        │      } 1 M
0C800H ├──────────────────┤
       │  DISK CACHE (6 K)│
       ├──────────────────┤
       │  BIT MAPPED      │
       │  GRAPHICS        │
       │  AREA            │
       │  (40 K)          │
02800H ├──────────────────┤
       │ CHARACTER FONT   │
       │     (8 K)        │
00800H ├──────────────────┤
00000H │POINTERS & VECTORS (2 K)│
       └──────────────────┘
```

Typical memory map for an 8-bit home computer

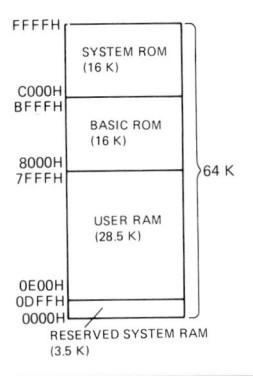

```
FFFFH ┌─────────────────┐
      │  SYSTEM ROM     │
      │  (16 K)         │
C000H │                 │
BFFFH ├─────────────────┤
      │  BASIC ROM      │
      │  (16 K)         │
8000H │                 │
7FFFH ├─────────────────┤  } 64 K
      │                 │
      │  USER RAM       │
      │  (28.5 K)       │
      │                 │
0E00H │                 │
0DFFH ├─────────────────┤
0000H └─────────────────┘
       RESERVED SYSTEM RAM
       (3.5 K)
```

Internal architecture of a microprocessor

The microprocessor forms the heart of any microcomputer system and thus its operation is crucial to the entire system. The principal constituents of a microprocessor are:

(a) registers for temporary storage of addresses, instructions, and data;
(b) an arithmetic logic unit (ALU) able to perform a variety of arithmetic and logical operations; and
(c) a control unit to provide control and timing signals for the entire system.

Internal architecture tends to vary widely; however, there are a few common themes. The major microprocessor families, for example, tend to retain a high degree of upward compatibility, both in terms of the internal architecture of the CPU and its major support devices and the instruction set employed. This is clearly an important consideration in making new products attractive to the equipment manufacturer.

Some of the CPU registers are directly accessible to the programmer while others are not. Registers may also be classified as either 'dedicated' or 'general purpose'. In the former case the register is reserved for a particular function such as pointing to a memory location or holding the result of an ALU operation.

The following CPU registers are worthy of special note:

Program Counter (PC)/Instruction Pointer (IP)

The program counter contains the address of the next instruction byte to be executed. An arguably better name for this register, and that adopted by Intel, is the Instruction Pointer.

The contents of the program counter or instruction pointer is automatically incremented by the CPU each time an instruction byte (or word) is fetched.

Accumulator (A)

The Accumulator (A) functions both as a source and a destination register; not only is it usually the source of one of the data bytes (or words) required for an ALU operation but it is also the location in which the result of the operation is placed.

Index Registers (I, X, Y, IX, IY etc.)

Index registers are normally used to facilitate operations on tables of data stored in memory. This is achieved by means of a 'base address' for the table, which is stored in the index register. The 'effective address' is then determined by adding an 'offset' or 'displacement' contained within a relevant indexed instruction.

Stack Pointer (SP)

Most CPUs need to have access to an area of external read/write memory (RAM) which facilitates temporary storage of data. The stack operates on a last-in, first-out (LIFO) basis. Data is 'pushed' onto the stack and later 'pulled' off it. The stack pointer contains the address in memory of the last used stack location.

Some processors provide two independent stack pointers. One is used to maintain the System Stack (S or SSP) while the other is available to control the User Stack (U or USP).

Flag Register (F)/Status Register (S)/Condition Code Register (CCR)

The flag register contains information on the current state of the microprocessor and, in particular, signals the result of the last ALU operation. The flag register is not a register in the conventional sense; it is simply a collection of bistables which are set or reset. Each bistable generates a signal which can be considered to be a 'flag'. Commonly available flags are: zero (Z), overflow (V), negative (N), and carry (C).

Simplified model of the internal architecture of a microprocessor

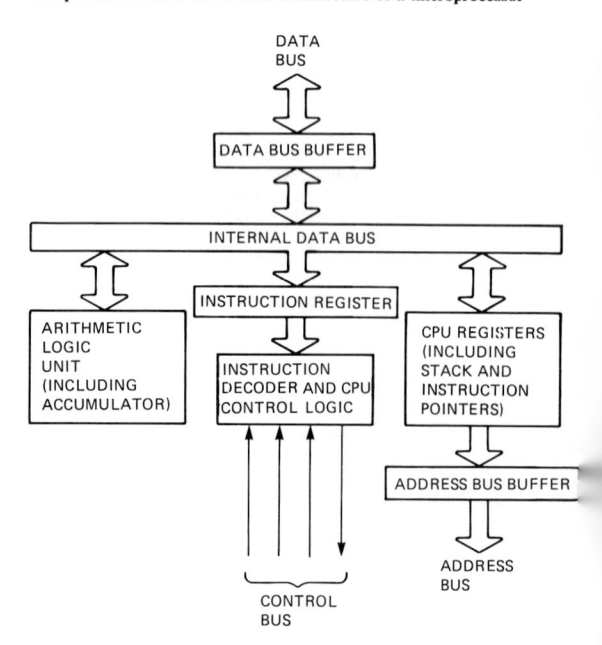

8-bit microprocessor architecture (8080 family)

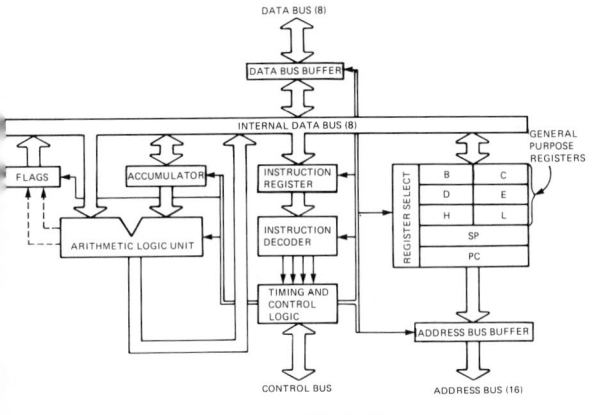

16-bit microprocessor architecture (8086 family)

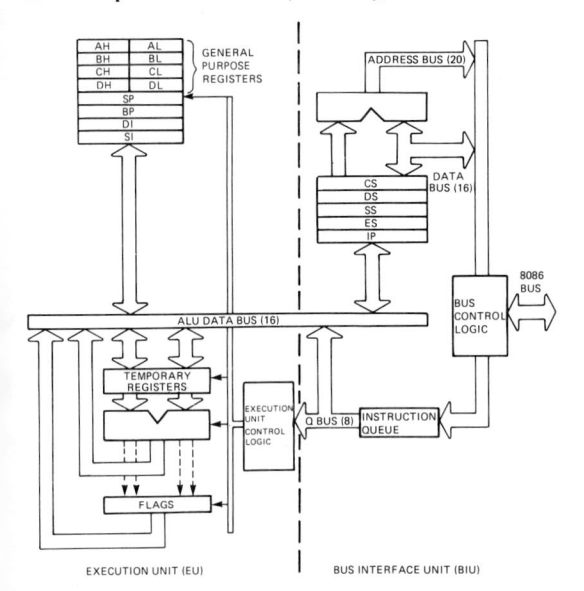

Some typical CPUs

6502
The 6502 was developed by MOS Technology as an improved 6800 device. The processor thus employs the same bus structure and is broadly compatible with the same range of peripheral support devices.

In the late 1970s and early 1980s the 6502 became widely accepted as an 'industry standard' 8-bit microprocessor, finding applications as varied as industrial process controllers and home computers.

The 6502 has only one general purpose data register (the accumulator), two 8-bit index registers (X and Y), an 8-bit stack pointer (S), and a 16-bit program counter (PC). An external single-phase clock is required, a typical frequency for which is 1 MHz. The device operates from a single +5 V supply.

The 6502 employs 13 addressing modes, 56 basic instructions, and has a total of 7 internal registers. Since the index registers are 8 rather than 16 bits wide, the 6502 requires a large number of addressing modes in order to achieve a full 64K indexed address range.

The following dedicated 6500 support devices are available:

6520 Parallel interface adaptor (PIA) with two 8-bit I/O ports
6522 Versatile interface adaptor (VIA) with two 8-bit I/O ports and two 16-bit timers
6532 RAM, I/O, Timer
6541 Keyboard/Display controller
6545 CRT controller
6551 Serial universal asynchronous receiver/transmitter (UART)

The 6502 is 'second sourced' by Rockwell and Synertek. CMOS versions of the 6502 family are now readily available, including 65C02 (CPU), 65C21 (PIA), 65C22 (VIA), 65C51 (ACIA), and 65C102 (enhanced CPU). These pin and instruction set compatible devices will operate at up to 2 MHz with vastly reduced power consumption.

Some typical CPUs 69

6502 basic configuration

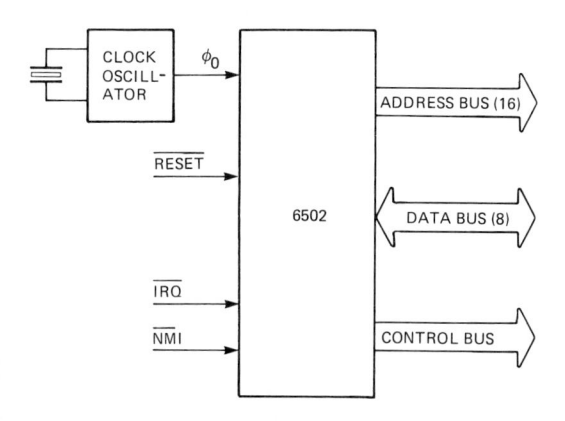

6502 register model

| ACCUMULATOR (A) |
| INDEX REGISTER (X) |
| INDEX REGISTER (Y) |
| PROGRAM COUNTER (PC) |
| STACK POINTER (S) |

6502 pin-out

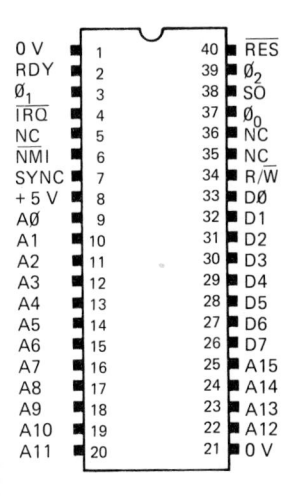

```
0 V      1       40  RES
RDY      2       39  Ø2
Ø1       3       38  SO
IRQ      4       37  Ø0
NC       5       36  NC
NMI      6       35  NC
SYNC     7       34  R/W
+5 V     8       33  D0
A0       9       32  D1
A1      10       31  D2
A2      11       30  D3
A3      12       29  D4
A4      13       28  D5
A5      14       27  D6
A6      15       26  D7
A7      16       25  A15
A8      17       24  A14
A9      18       23  A13
A10     19       22  A12
A11     20       21  0 V
```

6502 instruction set

MNEMONIC	OPERATION	IMMEDIATE OP c *	ABSOLUTE OP c *	ZERO PAGE OP c *	ACCUM OP c *	IMPLIED OP c *	(IND, X) OP c *	(IND), Y OP c *	Z PAGE, X OP c *	ABS, X OP c *	ABS, Y OP c *	RELATIVE OP c *	INDIRECT OP c *	Z PAGE, Y OP c *	PROCESSOR STATUS CODES 7 6 5 4 3 2 1 0 N V · B D I Z C
ADC	A + M + C → A (4)(1)	69 2 2	6D 4 3	65 3 2			61 6 2	71 5 2	75 4 2	7D 4 3	79 4 3				N V · · · · Z C
AND	A∧M → A (1)	29 2 2	2D 4 3	25 3 2			21 6 2	31 5 2	35 4 2	3D 4 3	39 4 3				N · · · · · Z ·
ASL	C←[7......0]←0		0E 6 3	06 5 2	0A 2 1				16 6 2	1E 7 3					N · · · · · Z C
BCC	BRANCH ON C = 0 (2)											90 2 2			· · · · · · · ·
BCS	BRANCH ON C = 1 (2)											B0 2 2			· · · · · · · ·
BEQ	BRANCH ON Z = 1 (2)											F0 2 2			· · · · · · · ·
BIT	A∧M, M₇→N, M₆→V		2C 4 3	24 3 2											M₇ M₆ · · · · Z ·
BMI	BRANCH ON N = 1 (2)											30 2 2			· · · · · · · ·
BNE	BRANCH ON Z = 0 (2)											D0 2 2			· · · · · · · ·
BPL	BRANCH ON N = 0 (2)											10 2 2			· · · · · · · ·
BRK	BREAK (See Fig 1)					00 7 1									· · · · · 0 · ·
BVC	BRANCH ON V = 0 (2)											50 2 2			· · · · · · · ·
BVS	BRANCH ON V = 1 (2)											70 2 2			· · · · · · · ·
CLC	0 → C					18 2 1									· · · · · · · 0
CLD	0 → D					D8 2 1									· · · · 0 · · ·
CLI	0 → I					58 2 1									· · · · · 0 · ·
CLV	0 → V					B8 2 1									· 0 · · · · · ·
CMP	A - M (1)	C9 2 2	CD 4 3	C5 3 2			C1 6 2	D1 5 2	D5 4 2	DD 4 3	D9 4 3				N · · · · · Z C
CPX	X - M	E0 2 2	EC 4 3	E4 3 2											N · · · · · Z C
CPY	Y - M	C0 2 2	CC 4 3	C4 3 2											N · · · · · Z C
DEC	M - 1 → M		CE 6 3	C6 5 2					D6 6 2	DE 7 3					N · · · · · Z ·
DEX	X - 1 → X					CA 2 1									N · · · · · Z ·
DEY	Y - 1 → Y					88 2 1									N · · · · · Z ·
EOR	A∀M → A (1)	49 2 2	4D 4 3	45 3 2			41 6 2	51 5 2	55 4 2	5D 4 3	59 4 3				N · · · · · Z ·
INC	M + 1 → M		EE 6 3	E6 5 2					F6 6 2	FE 7 3					N · · · · · Z ·
INX	X + 1 → X					E8 2 1									N · · · · · Z ·
INY	Y + 1 → Y					C8 2 1									N · · · · · Z ·
JMP	JUMP TO NEW LOC		4C 3 3										6C 5 3		· · · · · · · ·
JSR	JUMP SUB (See Fig 2)		20 6 3												· · · · · · · ·
LDA	M → A (1)	A9 2 2	AD 4 3	A5 3 2			A1 6 2	B1 5 2	B5 4 2	BD 4 3	B9 4 3				N · · · · · Z ·

Some typical CPUs 71

L D X	M → X	(1)	A2 2 2	AE 4 3	A6 3 2		B6 4 2		BE 4 3											N · · · Z		
L D Y	M → Y	(1)	A0 2 2	AC 4 3	A4 3 2	2 2 4A			BC 4 3											N · · · Z		
L S R	0 → □ → C			4E 6 3	46 5 2		56 6 2		5E 7 3											0 · · · Z C		
N O P	NO OPERATION							EA 2 1													· · · · · ·		
O R A	A V M → A		09 2 2	0D 4 3	05 3 2		15 4 2	01 6 2	1D 4 3		19 4 3		11 5 2							N · · · Z		
P H A	A → Ms															48 3 1					· · · · · ·		
P H P	P → Ms															08 3 1					· · · · · ·		
P L A	S + 1 → S, Ms → A															68 4 1					N · · · Z		
P L P	S + 1 → S, Ms → P															28 4 1					(RESTORED)		
R O L	□ → □			2E 6 3	26 5 2		36 6 2		3E 7 3							2A 2 1					N · · · Z C		
R O R	□ → □			6E 6 3	66 5 2		76 6 2		7E 7 3							6A 2 1					N · · · Z C		
R T I	RTRN INT															40 6 1					(RESTORED)		
R T S	RTRN SUB															60 6 1					· · · · · ·		
S B C	A − M − C → A		E9 2 2	ED 4 3	E5 3 2		F5 4 2	E1 6 2	FD 4 3		F9 4 3		F1 5 2							N V · · Z C		
S E C	1 → C															38 2 1					· · · · · C		
S E D	1 → D															F8 2 1					· · · · D ·		
S E I	1 → I															78 2 1					· · I · · ·		
S T A	A → M			8D 4 3	85 3 2		95 4 2	81 6 2	9D 5 3		99 5 3		91 6 2							· · · · · ·		
S T X	X → M			8E 4 3	86 3 2		96 4 2														· · · · · ·		
S T Y	Y → M			8C 4 3	84 3 2		94 4 2														· · · · · ·		
T A X	A → X															AA 2 1					N · · · Z		
T A Y	A → Y															A8 2 1					N · · · Z		
T S X	S → X															BA 2 1					N · · · Z		
T X A	X → A															8A 2 1					N · · · Z		
T X S	X → S															9A 2 1					· · · · · ·		
T Y A	Y → A															98 2 1					N · · · Z		

(1) ADD 1 to 'N' IF PAGE BOUNDARY IS CROSSED
(2) ADD 1 TO 'N' IF BRANCH OCCURS TO SAME PAGE
    ADD 2 TO 'N' IF BRANCH OCCURS TO DIFFERENT PAGE
(3) CARRY NOT = BORROW
(4) IF IN DECIMAL MODE Z FLAG IS INVALID
    ACCUMULATOR MUST BE CHECKED FOR ZERO RESULT

X INDEX X
Y INDEX Y
A ACCUMULATOR
M MEMORY PER EFFECTIVE ADDRESS
Ms MEMORY PER STACK POINTER

+ ADD
− SUBTRACT
∧ AND
∨ OR
∀ EXCLUSIVE OR

M₇ MEMORY BIT 7
M₆ MEMORY BIT 6
n   NO. CYCLES
NO. BYTES

6809

The 6809 was developed by Motorola as a successor to the 6800 device and as a rival to the 6502. The device uses an expanded 6800-type instruction set but incorporates a number of 16-bit instructions. To some extent, therefore, the 6809 forms a bridge between the second generation of 8-bit processors and the first generation of 16-bit processors.

The 6809 arrived rather too late to effectively rival the 6502 and Z80 devices in the mass production of 8-bit microcomputers, with only two manufacturers (Dragon Data and Tandy) adopting the device for the home computer boom of the early 1980s. The 6809 has, however, gained a large following in the industrial control sector where its power and elegance make it an attractive alternative to the 6502, 6800 and Z80. For this reason alone, the 6809 is likely to maintain its position as a versatile work-horse for many years to come.

The 6809 has two separate 8-bit accumulators (A and B). These may be used together to form a single 16-bit accumulator for use in 16-bit operations. Two 16-bit index registers (X and Y) are available, as are a 16-bit user stack pointer (U), 16-bit system stack pointer (S), and 16-bit program counter (PC).

The 6809 instruction set has 59 basic instructions including nine 16-bit accumulator/memory operations.

Typical 6800 family support devices include:

6821 Peripheral interface adaptor (PIA)
6828 Priority interrupt controller
6840 Programmable timer module
6845 CRT controller
6850 Asynchronous communications interface adaptor (ACIA)
6852 Synchronous serial data adaptor (SSDA)

The 6809 is 'second sourced' by AMI, Fairchild, and Hitachi. Since the device has retained bus compatibility with the 6800 and 6502 devices, a very wide range of support devices is available.

A CMOS version of the 6809 (HD6309) is available from Hitachi. The device operates with clock frequencies over the range 500 kHz to 8 MHz with a supply of 5 V $\pm 10\%$ and features vastly reduced power consumption.

6809 basic configuration

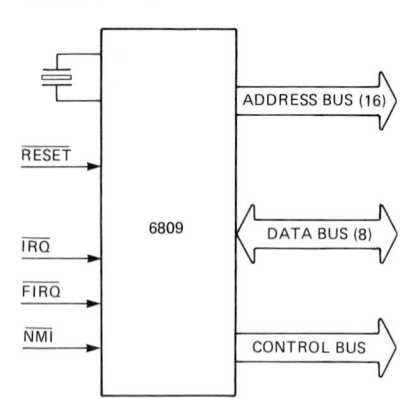

6809 register model

ACCUMULATOR (A)	ACCUMULATOR (B)
DIRECT PAGE (DP)	FLAGS (CC)
INDEX REGISTER (X)	
INDEX REGISTER (Y)	
USER STACK POINTER (U)	
SYSTEM STACK POINTER (S)	
PROGRAM COUNTER (PC)	

6809 pinout

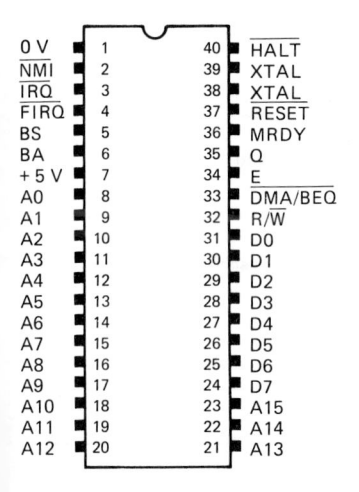

```
 0 V    1        40  HALT
 NMI    2        39  XTAL
 IRQ    3        38  XTAL
 FIRQ   4        37  RESET
 BS     5        36  MRDY
 BA     6        35  Q
 +5 V   7        34  E
 A0     8        33  DMA/BEQ
 A1     9        32  R/W
 A2    10        31  D0
 A3    11        30  D1
 A4    12        29  D2
 A5    13        28  D3
 A6    14        27  D4
 A7    15        26  D5
 A8    16        25  D6
 A9    17        24  D7
 A10   18        23  A15
 A11   19        22  A14
 A12   20        21  A13
```

6809 instruction set

8-BIT ACCUMULATOR AND MEMORY

Mnemonic(s)	Operation
ADCA, ADCB	Add memory to accumulator with carry
ADDA, ADDB	Add memory to accumulator
ANDA, ANDB	Add memory with accumulator
ASL, ALSA, ASLB	Arithmetic shift of accumulator or memory left
ASR, ASRA, ASRB	Arithmetic shift of accumulator or memory right
BITA, BITB	Bit test memory with accumulator
CLR, CLRA, CLRB	Clear accumulator or memory location
CMPA, CMPB	Compare memory from accumulator
COM, COMA, COMB	Complement accumulator or memory location
DAA	Decimal adjust A-accumulator

Some typical CPUs

Mnemonic(s)	Operation
DEC, DECA, DECB	Decrement accumulator or memory location
EORA, EORB	Exclusive or memory with accumulator
EXG R1, R2	Exchange R1 with R2 (R1, R2 = A, B, CC, DP)
INC, INCA, INCB	Increment accumulator or memory location
LDA, LDB	Load accumulator from memory
LSL, LSLA, LSLB	Logical shift left accumulator or memory location
LSR, LSRA, LSRB	Logical shift right accumulator or memory location
MUL	Unsigned multiply (A x B→D)
NEG, NEGA, NEGB	Negate accumulator or memory
ORA, ORB	Or memory with accumulator
ROL, ROLA, ROLB	Rotate accumulator or memory left
ROR, RORA, RORB	Rotate accumulator or memory right
SBCA, SBCB	Subtract memory from accumulator with borrow
STA, STB	Store accumulator to memory
SUBA, SUBB	Subtract memory from accumulator
TST, TSTA, TSTB	Test accumulator or memory location
TFR, R1, R2	Transfer R1 to R2 (R1, R2 = A, B, CC, DP)

NOTE: A, B, CC, or DP may be pushed to (pulled from) either stack with PSHS, PSHU, (PULS, PULU) instructions

16-BIT ACCUMULATOR AND MEMORY

Mnemonic(s)	Operation
ADDD	Add memory to D accumulator
CMPD	Compare memory from D accumulator
EXG D, R	Exchange D with X, Y, S, U or PC
LDD	Load D accumulator from memory
SEX	Sign Extend B accumulator into A accumulator
STD	Store D accumulator to memory
SUBD	Subtract memory from D accumulator
TFR D, R	Transfer D to X, Y, S, U or PC
TFR R, D	Transfer X, Y, S, U or PC to D

INDEX REGISTER STACK POINTER

Mnemonic(s)	Operation
CMPS, CMPU	Compare memory from stack pointer
CMPX, CMPY	Compare memory from index register
EXG R1, R2	Exchange D, X, Y, S, U or PC with D, X, S, Y, U or PC

Some typical CPUs 75

LEAS, LEAU	Load effective address into stack pointer
LEAX, LEAY	Load effective address into index register
LDS, LDU	Load stack pointer from memory
LDX, LDY	Load index register from memory
PSHS	Push any register(s) onto hardware stack (except S)
PSHU	Push any register(s) onto user stack (except U)
PULS	Pull any register(s) from hardware stack (except S)
PULU	Pull any register(s) from hardware stack (except U)
STS, STU	Store stack pointer to memory
STX, STY	Store index register to memory
TFR R1, R2	Transfer D, X, Y, S, U or PC to D, X, Y, S, U or PC
ABX	Add B accumulator to X (unsigned)

BRANCH

BCC, LBCC	Branch if carry clear
BCS, LBCS	Branch if carry set
BEQ, LBEQ	Branch if equal
BGE, LBGE	Branch if greater than or equal (signed)
BGT, LBGT	Branch if greater (signed)
BHI, LBHI	Branch if higher (unsigned)
BHS, LBHS	Branch if higher or same (unsigned)
BLE, LBLE	Branch if less than or equal (signed)
BLO, LBLO	Branch if lower (unsigned)
BLS, LBLS	Branch if lower or same (unsigned)
BLT, LBLT	Branch if less than (signed)
BMI, LBMI	Branch if minus
BNE, LBNE	Branch if not equal
BPL, LBPL	Branch if plus
BRA, LBRA	Branch always
BRN, LBRN	Branch never
BSR, LBSR	Branch to subroutine
BVC, LBVC	Branch if overflow clear
BVS, LBVS	Branch if overflow set

MISCELLANEOUS

ANDCC	AND condition code register
CWAI	AND condition code register, then wait for interrupt
NOP	No operation
ORCC	OR condition code register
JMP	Jump
JSR	Jump to subroutine
RTI	Return from interrupt
RTS	Return from subroutine
SW1, SW12, SW13	Software interrupt (absolute indirect)
SYNC	Synchronise with interrupt line

Z80

The Z80 is a powerful 8-bit microprocessor which has a total of 158 basic instructions (including bit set and test, and block move). The Z80 was designed by Zilog as a very much enhanced 8080 device and yet retains full instruction set compatibility with that device.

The Z80 has 17 internal registers including a duplicate set of general purpose registers. Two 16-bit index registers (IX and IY) are provided as well as a 16-bit stack pointer (SP) and 16-bit program counter (PC). The device also provides for three interrupt modes.

An interesting feature of the Z80 (and one which earned it considerable popularity with manufacturers) is the internal provision for refreshing dynamic RAM.

For control applications, the Z80 provides IN and OUT instructions which permit reading and writing data to any one of 256 I/O ports. Memory and port addresses are distinguished using the memory request (MREQ) and input/output request (IORQ) lines.

The Z80 operates from a single +5 V supply and requires a single-phase clock at typical frequencies of 4 MHz (Z80A) or 6 MHz (Z80B). It should also be noted that the clock input to the Z80 normally requires a 390 Ω pull-up resistor.

Alternative sources for the Z80 include Fairchild (F3880), Mostek (MK3880), NEC (μPD780C), and SGS-ATES (Z80A).

Z80 support devices include:

Z80-CTC (Z8430)	Counter/timer circuit
Z80-DART (Z8470)	Dual asynchronous receiver/transmitter
Z80-DMA (Z8410)	Direct memory access controller
Z80-PIO (Z8420)	Parallel input/output controller
Z80-SIO/0 (Z8440)	Serial input/output controller (with two synchronizing inputs)
Z80-SIO/2 (Z8442)	Serial input/output controller (with one synchronizing input and independent receiver and transmitter)

Support devices are coded with a suffix: A to indicate 4 MHz operation, and B to indicate 6 MHz operation.

Z80 basic configuration

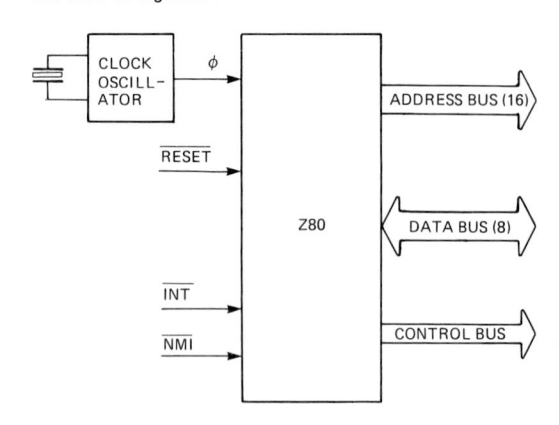

Some typical CPUs 77

A CMOS version of the Z80 is available from NEC. This μPD7000PC device operates at up to 4 MHz with supplies of between 3 V and 6 V. Typical operating and standby currents are 16 mA and 100 μA respectively.

National Semiconductor produce a CMOS Z80-compatible device (the NSC800) which uses the full Z80 instruction set. The device is available in 1 MHz (NSC800N-1), 2.5 MHz (NSC800N), and 4 MHz (NSC800N-4) versions and it operates over a supply voltage range of 3 V to 12 V.

CMOS support devices (Z84C20 and Z84C30) are available from Toshiba, as is a Z80 CMOS clock generator (6497).

Z80 register model

MAIN REGISTER SET

ACCUMULATOR (A)	FLAGS	(F)
(B)		(C)
(D)		(E)
(H)		(L)

ALTERNATE REGISTER SET

ACCUMULATOR (A')	FLAGS	(F')
(B')		(C')
(D')		(E')
(H')		(L')

SPECIAL PURPOSE REGISTERS

INTERRUPT VECTOR (I)	MEMORY REFRESH (R)
INDEX REGISTER	(IX)
INDEX REGISTER	(IY)
STACK POINTER	(SP)
PROGRAM COUNTER	(PC)

Z80 pinout

Pin			Pin	
A11	1	40	A10
A12	2	39	A9
A13	3	38	A8
A14	4	37	A7
A15	5	36	A6
Ø	6	35	A5
D4	7	34	A4
D3	8	33	A3
D5	9	32	A2
D6	10	31	A1
+5 V	11	30	A0
D2	12	29	OV
D7	13	28	RFSH
D0	14	27	M1
D1	15	26	RESET
INT	16	25	BUSRQ
NMI	17	24	WAIT
HALT	18	23	BUSAK
MREQ	19	22	WR
IORQ	20	21	RD

Z80 instruction set

	Mnemonic	Symbolic Operation	Comments
8-BIT LOADS	LD r, s	r ← s	s ≡ r, n, (HL), (IX+e), (IY+e)
	LD d, r	d ← r	d ≡ (HL), r (IX+e), (IY+e)
	LD d, n	d ← n	d ≡ (HL), (IX+e), (IY+e)
	LD A, s	A ← s	s ≡ (BC), (DE), (nn), I, R
	LD d, A	d ← A	d ≡ (BC), (DE), (nn), I, R
16-BIT LOADS	LD dd, nn	dd ← nn	dd ≡ BC, DE, HL, SP, IX, IY
	LD dd, (nn)	dd ← (nn)	dd ≡ BC, DE, HL, SP, IX, IY
	LD (nn), ss	(nn) ← ss	ss ≡ BC, DE, HL, SP, IX, IY
	LD SP, ss	SP ← ss	ss ≡ HL, IX, IY
	PUSH ss	(SP-1) ← ss_H; (SP-2) ← ss_L	ss ≡ BC, DE, HL, AF, IX, IY
	POP dd	dd_L ← (SP); dd_H ← (SP+1)	dd ≡ BC, DE, HL, AF, IX, IY
EXCHANGES	EX DE, HL	DE ↔ HL	
	EX-AF, AF'	AF ↔ AF'	
	EXX	$\begin{pmatrix}BC\\DE\\HL\end{pmatrix} \leftrightarrow \begin{pmatrix}BC'\\DE'\\HL'\end{pmatrix}$	
	EX (SP), ss	(SP) ↔ ss_L, (SP+1) ↔ ss_H	ss ≡ HL, IX, IY
MEMORY BLOCK MOVES	LDI	(DE) ← (HL), DE ← DE+1 HL ← HL+1, BC ← BC−1	
	LDIR	(DE) ← (HL), DE ← DE+1 HL ← HL+1, BC ← BC−1 Repeat until BC = 0	
	LDD	(DE) ← (HL), DE ← DE−1 HL ← HL−1, BC ← BC−1	
	LDDR	(DE) ← (HL), DE ← DE−1 HL ← HL−1, BC ← BC−1 Repeat until BC = 0	
MEMORY BLOCK SEARCHES	CPI	A−(HL), HL ← HL+1 BC ← BC−1	A−(HL) sets the flags only. A is not affected
	CPIR	A−(HL), HL ← HL+1 BC ← BC−1, Repeat until BC = 0 or A = (HL)	
	CPD	A−(HL), HL ← HL−1 BC ← BC−1	
	CPDR	A−(HL), HL ← HL−1 BC ← BC−1, Repeat until BC=0 or A=(HL)	
8-BIT ALU	ADD s	A ← A + s	CY is the carry flag
	ADC s	A ← A + s + CY	
	SUB s	A ← A − s	
	SBC s	A ← A − s − CY	s ≡ r, n, (HL) (IX+e), (IY+e)
	AND s	A ← A ∧ s	
	OR s	A ← A ∨ s	
	XOR s	A ← A ⊕ s	
	CP s	A − s	s ≡ r, n (HL) (IX+e), (IY+e)
	INC d	d ← d + 1	d ≡ r, (HL) (IX+e), (IY+e)
	DEC d	d ← d − 1	
	ADD HL, ss	HL ← HL + ss	ss ≡ BC, DE HL, SP
	ADC HL, ss	HL ← HL + ss + CY	
	SBC HL, ss	HL ← HL − ss − CY	
	ADD IX, ss	IX ← IX + ss	ss ≡ BC, DE, IX, SP

Some typical CPUs

	Mnemonic	Symbolic Operation	Comments
GP 16-BIT ARITHMETIC	ADD IY, ss	IY ← IY + ss	ss ≡ BC, DE, IY, SP
	INC dd	dd ← dd + 1	dd ≡ BC, DE, HL, SP, IX, IY
	DEC dd	dd ← dd − 1	dd ≡ BC, DE, HL, SP, IX, IY
GP ACC. & FLAG	DAA	Converts A contents into packed BCD following add or subtract.	Operands must be in packed BCD format
	CPL	A ← $\overline{A}$	
	NEG	A ← 00 − A	
	CCF	CY ← $\overline{CY}$	
	SCF	CY ← 1	
MISCELLANEOUS	NOP	No operation	
	HALT	Halt CPU	
	DI	Disable Interrupts	
	EI	Enable Interrupts	
	IM 0	Set interrupt mode 0	8080A mode
	IM 1	Set interrupt mode 1	Call to 0038$_H$
	IM 2	Set interrupt mode 2	Indirect Call
ROTATES AND SHIFTS	RLC s		s ≡ r, (HL) (IX+e), (IY+e)
	RL s		
	RRC s		
	RR s		
	SLA s		
	SRA s		
	SRL s		
	RLD		
	RRD		
BIT S, R, & T	BIT b, s	Z ← $\overline{s_b}$	Z is zero flag
	SET b, s	s_b ← 1	s ≡ r, (HL)
	RES b, s	s_b ← 0	(IX+e), (IY+e)
INPUT AND OUTPUT	IN A, (n)	A ← (n)	
	IN r, (C)	r ← (C)	Set flags
	INI	(HL) ← (C), HL ← HL + 1 B ← B − 1	
	INIR	(HL) ← (C), HL ← HL + 1 B ← B − 1 Repeat until B = 0	
	IND	(HL) ← (C), HL ← HL − 1 B ← B − 1	
	INDR	(HL) ← (C), HL ← HL − 1 B ← B − 1 Repeat until B = 0	
	OUT(n), A	(n) ← A	
	OUT(C), r	(C) ← r	
	OUTI	(C) ← (HL), HL ← HL + 1 B ← B − 1	
	OTIR	(C) ← (HL), HL ← HL + 1 B ← B − 1 Repeat until B = 0	
	OUTD	(C) ← (HL), HL ← HL − 1 B ← B − 1	
	OTDR	(C) ← (HL), HL ← HL − 1 B ← B − 1 Repeat until B = 0	

80 Some typical CPUs

JUMPS	JP nn	PC ← nn		
	JP cc, nn	If condition cc is true PC ← nn, else continue	cc	NZ PO Z PE NC P C M
	JR e	PC ← PC + e		
	JR kk, e	If condition kk is true PC ← PC + e, else continue	kk	NZ NC Z C
	JP (ss)	PC ← ss	ss = HL, IX, IY	
	DJNZ e	B ← B − 1, if B = 0 continue, else PC · PC + e		

CALLS	CALL nn	(SP−1) ← PC$_H$ (SP−2) ← PC$_L$, PC ← nn		
	CALL cc, nn	If condition cc is false continue, else same as CALL nn	cc	NZ PO Z PE NC P C M

RESTARTS	RST L	(SP−1) ← PC$_H$ (SP−2) ← PC$_L$, PC$_H$ ← 0 PC$_L$ ← L

RETURNS	RET	PC$_L$ ← (SP), PC$_H$ ← (SP+1)		
	RET cc	If condition cc is false continue, else same as RET	cc	NZ PO Z PE NC P C M
	RETI	Return from interrupt, same as RET		
	RETN	Return from non- maskable interrupt		

In the table the following abbreviations are used.
b ≡ a bit number in any 8-bit register or memory location
cc ≡ flag condition code

NZ	≡ non zero		PO	≡ Parity odd or no over flow
Z	≡ zero		PE	≡ Parity even or over flow
NC	≡ non carry		P	≡ Positive
C	≡ carry		M	≡ Negative (minus)

d ≡ any 8-bit destination register or memory location
dd ≡ any 16-bit destination register or memory location
e ≡ 8-bit signed 2's complement displacement used in relative jumps and indexed addressing
L ≡ 8 special call locations in page zero. In decimal notation these are 0, 8, 16, 24, 32, 40, 48 and 56
n ≡ any 8-bit binary number
nn ≡ any 16-bit binary number
r ≡ any 8-bit general purpose register (A, B, C, D, E, H, or L)
s ≡ any 8-bit source register or memory location
s$_b$ ≡ a bit in a specific 8-bit register or memory location
ss ≡ any 16-bit source register or memory location
subscript "L" ≡ the low order 8 bits of a 16-bit register
subscript "H" ≡ the high order 8 bits of a 16-bit register
( ) ≡ the contents within the ( ) are to be used as a pointer to a memory location or I/O port number
8-bit registers are A, B, C, D, E, H, L, I and R
16-bit register pairs are AF, BC, DE and HL
16-bit registers are SP, PC, IX and IY

8086

The Intel 8086 was the first true second-generation 16-bit microprocessor. Its arrival was extremely timely and coincided with the availability of low-cost high-capacity semiconductor memories. The 8086 thus rapidly became the first industry-standard 16-bit microprocessor. The 8086 instruction set has retained some compatibility with the 8-bit 8080 instructions and is well suited to efficient compilation from such high level languages as BASIC and Pascal.

The 8086 has 14 16-bit registers (including those that are the direct equivalents of 8080/8085 registers). The CPU has 20 address lines and thus provides for a 1M byte address range. The I/O address range, on the other hand, is 64K bytes.

Like many of its rival 16-bit processors, the 8086 uses a segmented address system. Four segment registers are provided: code segment (CS), stack segment (SS), data segment (DS) and extra segment (ES). A physical segment must start on a 16-byte address boundary (the four least significant address bits are all set to zero) and have a size of 64K bytes.

The actual 20-bit address is formed by extending the 16-bit segment address by shifting and adding four least significant zero bits (i.e. effectively multiplying the segment register contents by 16). The contents of the instruction pointer (IP), stack pointer (SP) or other 16-bit address register are then added to the result.

The 8086 operates from a single +5 V supply with typical clock frequencies of 5 MHz, 8 MHz and 10 MHz.

In order to avoid the use of a large package, the 8086 uses a multiplexed address/data bus; the lower 16 address lines share the 16 data bus lines.

The device can operate in one of two modes, 'max' or 'min'. Maximum mode is employed in multi-processor systems where an 8288 bus controller device decodes the 8086 S0, S1, and S2 lines to produce the necessary I/O and memory control signals.

Dedicated 8086 support devices include:

8087	Numeric data co-processor
8089	I/O processor with DMA
8207	Dynamic RAM controller
8208	Dynamic RAM controller
8237	DMA controller
8253	Timer/counter
8254	Timer/counter
8259A	Interrupt controller
8284A	Clock generator (essential for 8086-based systems)
8286	Data bus transceiver
8287	Data bus transceiver
8288	Bus controller (essential for 'max' mode)

The 8086 can also make use of the 8080 family of support devices.

The 8086 is 'second sourced' by AMD, Fujitsu, NEC, OKI, and Siemens. CMOS versions (80C86) are available from Harris and OKI.

8086 basic configuration

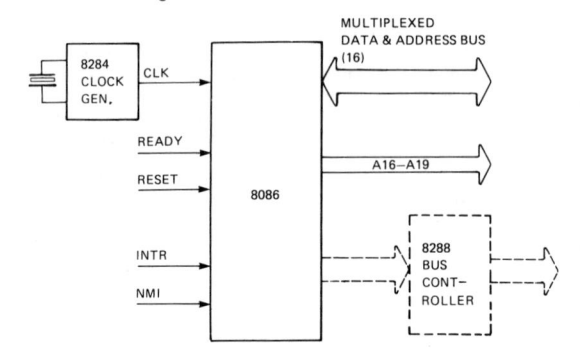

8086 register model

POINTER AND INDEX REGISTERS	
STACK POINTER	(SP)
BASE POINTER	(BP)
SOURCE INDEX	(SI)
DESTINATION INDEX	(DI)

GENERAL PURPOSE REGISTERS			
(AH)	ACC.	(AX)	(AL)
(BH)	BASE	(BX)	(BL)
(CH)	COUNT	(CX)	(CL)
(DH)	DATA	(DX)	(DL)

SEGMENT REGISTERS	
CODE SEGMENT	(CS)
DATA SEGMENT	(DS)
STACK SEGMENT	(SS)
EXTRA SEGMENT	(ES)

INSTRUCTION POINTER	(IP)
STATUS	(ST)

8086 pinout

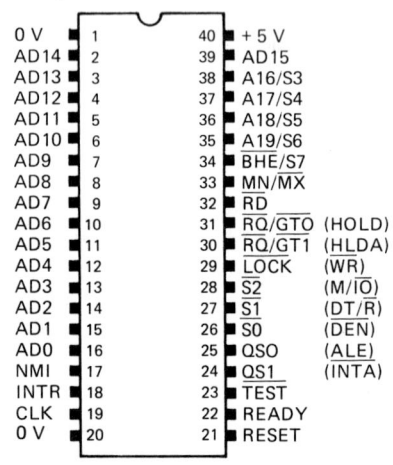

```
0 V        1      40    +5 V
AD14       2      39    AD15
AD13       3      38    A16/S3
AD12       4      37    A17/S4
AD11       5      36    A18/S5
AD10       6      35    A19/S6
AD9        7      34    BHE/S7
AD8        8      33    MN/MX
AD7        9      32    RD
AD6       10      31    RQ/GT0  (HOLD)
AD5       11      30    RQ/GT1  (HLDA)
AD4       12      29    LOCK    (WR)
AD3       13      28    S2      (M/IO)
AD2       14      27    S1      (DT/R)
AD1       15      26    S0      (DEN)
AD0       16      25    QS0     (ALE)
NMI       17      24    QS1     (INTA)
INTR      18      23    TEST
CLK       19      22    READY
0 V       20      21    RESET
```

Note: Signals shown in brackets correspond to 'minimum' mode when MN/$\overline{\text{MX}}$ is high.

Some typical CPUs 83

8086 instruction set

GENERAL PURPOSE

MOV	Move byte or word
PUSH	Push word onto stack
POP	Pop word off stack
PUSHA	Push all registers on stack
POPA	Pop all registers from stack
XCHG	Exchange byte or word
XLAT	Translate byte

INPUT/OUTPUT

IN	Input byte or word
OUT	Output byte or word

ADDRESS OBJECT

LEA	Load effective address
LDS	Load pointer using DS
LES	Load pointer using ES

FLAG TRANSFER

LAHF	Load AH register from flags
SAHF	Store AH register in flags
PUSHF	Push flags onto stack
POPF	Pop flags off stack

ADDITION

ADD	Add byte or word
ADC	Add byte or word with carry
INC	Increment byte or word by 1
AAA	ASCII adjust for addition
DAA	Decimal adjust for addition

SUBTRACTION

SUB	Subtract byte or word
SBB	Subtract byte or word with borrow
DEC	Decrement byte or word by 1
NEG	Negate byte or word
CMP	Compare byte or word
AAS	ASCII adjust for subtraction
DAS	Decimal adjust for subtraction

MULTIPLICATION

MUL	Multiply byte or word unsigned
IMUL	Integer multiply byte or word
AAM	ASCII adjust for multiply

DIVISION

DIV	Divide byte or word unsigned
IDIV	Integer divide byte or word
AAD	ASCII adjust for division
CBW	Convert byte to word
CWD	Convert word to doubleword

MOVS	Move byte or word string
INS	Input bytes or word string
OUTS	Output bytes or word string
CMPS	Compare byte or word string
SCAS	Scan byte or word string
LODS	Load byte or word string
STOS	Store byte or word string
REP	Repeat
REPE/REPZ	Repeat while equal zero
REPNE/REPNZ	Repeat while not equal not zero

LOGICALS

NOT	`Not` byte or word
AND	`And` byte or word
OR	`Inclusive or` byte or word
XOR	`Exclusive or` byte or word
TEST	`Test` byte or word

SHIFTS

SHL/SAL	Shift logical arithmetic left byte or word
SHR	Shift logical right byte or word
SAR	Shift arithmetic right byte or word

ROTATES

ROL	Rotate left byte or word
ROR	Rotate right byte or word
RCL	Rotate through carry left byte or word
RCR	Rotate through carry right byte or word

FLAG OPERATIONS

STC	Set carry flag
CLC	Clear carry flag
CMC	Complement carry flag
STD	Set direction flag
CLD	Clear direction flag
STI	Set interrupt enable flag
CLI	Clear interrupt enable flag

EXTERNAL SYNCHRONIZATION

HLT	Halt until interrupt or reset
WAIT	Wait for TEST pin active
ESC	Escape to extension processor
LOCK	Lock bus during next instruction

NO OPERATION

NOP	No operation

HIGH LEVEL INSTRUCTIONS

ENTER	Format stack for procedure entry
LEAVE	Restore stack for procedure exit
BOUND	Detects values outside prescribed range

CONDITIONAL TRANSFERS

JA/JNBE	Jump if above not below nor equal
JAE/JNB	Jump if above or equal not below
JB/JNAE	Jump if below not above nor equal
JBE/JNA	Jump if below or equal not above
JC	Jump if carry
JE/JZ	Jump if equal/zero
JG/JNLE	Jump if greater/not less nor equal
JGE/JNL	Jump if greater or equal/not less
JL/JNGE	Jump if less/not greater nor equal
JLE/JNG	Jump if less or equal not greater
JNC	Jump if not carry
JNE/JNZ	Jump if not equal/not zero
JNO	Jump if not overflow
JNP/JPO	Jump if not parity/parity odd
JNS	Jump if not sign
JO	Jump if overflow
JP/JPE	Jump if parity/parity even
JS	Jump if sign

UNCONDITIONAL TRANSFERS

CALL	Call procedure
RET	Return from procedure
JMP	Jump

ITERATION CONTROLS

LOOP	Loop
LOOPE/LOOPZ	Loop if equal/zero
LOOPNE/LOOPNZ	Loop if not equal/not zero
JCXZ	Jump if register CX 0

INTERRUPTS

INT	Interrupt
INTO	Interrupt if overflow
IRET	Interrupt return

8088

The 8088 is an 8-bit data bus version of the 8086. The device has the same internal architecture as the 8086 and shares the same instruction set. This permits full software migration between the two systems and the same range of support devices.

The 8088 attracted much support from equipment manufacturers wishing to produce 16-bit microcomputers while retaining lower cost 8-bit data bus systems. The 8088 was thus adopted for the IBM PC, ACT Sirius, and Sanyo MBC-555 to name just three.

For other details see the 8086.

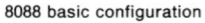

8088 basic configuration

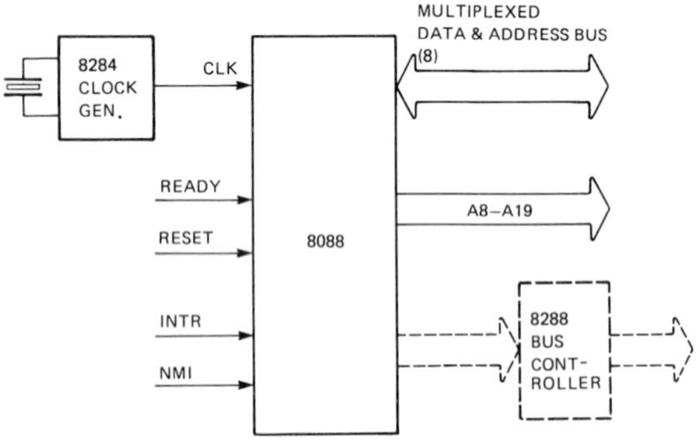

8088 register model

POINTER AND INDEX REGISTERS

STACK POINTER (SP)
BASE POINTER (BP)
SOURCE INDEX (SI)
DESTINATION INDEX (DI)

GENERAL PURPOSE REGISTERS

(AH)	ACC	(AX)	(AL)
(BH)	BASE	(BX)	(BL)
(CH)	COUNT	(CX)	(CL)
(DH)	DATA	(DX)	(DL)

SEGMENT REGISTERS

CODE SEGMENT	(CS)
DATA SEGMENT	(DS)
STACK SEGMENT	(SS)
EXTRA SEGMENT	(ES)

INSTRUCTION POINTER	(IP)
STATUS	(ST)

8088 pinout

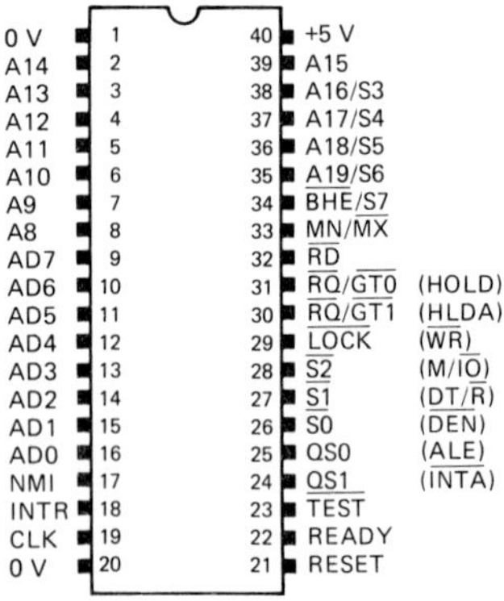

```
0 V      1    40   +5 V
A14      2    39   A15
A13      3    38   A16/S3
A12      4    37   A17/S4
A11      5    36   A18/S5
A10      6    35   A19/S6
A9       7    34   BHE/S7
A8       8    33   MN/MX
AD7      9    32   RD
AD6     10    31   RQ/GT0   (HOLD)
AD5     11    30   RQ/GT1   (HLDA)
AD4     12    29   LOCK     (WR)
AD3     13    28   S2       (M/IO)
AD2     14    27   S1       (DT/R)
AD1     15    26   S0       (DEN)
AD0     16    25   QS0      (ALE)
NMI     17    24   QS1      (INTA)
INTR    18    23   TEST
CLK     19    22   READY
0 V     20    21   RESET
```

Note: Signals shown in brackets correspond to 'minimum' mode when MN/$\overline{\text{MX}}$ is high.

68000

The Motorola 68000 is an advanced 16-bit microprocessor which features 32-bit internal architecture. The 68000 contains a total of 32 registers, uses a 24-bit address and data buses respectively, and offers a 16M byte direct addressing range. The device is housed in a 64-pin DIL package.

The internal architecture of the 68000 is delightfully 'clean'; it has eight 32-bit data registers (somewhat confusingly known as D0 to D7) and seven 32-bit address registers (A0 to A6). In addition, there are two 32-bit stack pointers (USP and SSP) and one 32-bit program counter (PC).

The 68000 requires a single-phase clock input at either 8 MHz (68000-8) or 12 MHz (68000-12) and operates from a single +5 V supply.

The 68000 features 56 basic instructions. Operations may be based on bits, bytes (8 bits), words (16 bits), and long-words (32 bits). The CPU can operate in 'user' or 'supervisor' modes and a further 'trace' mode is provided which will generate an exception (trap) every time an instruction is executed.

The 68000 features asynchronous bus transfers; an acknowledge signal is required for data transfers on the bus. The 68000 is compatible with the range of 8-bit 68000 support devices. Dedicated 68000 peripheral devices include:

- 68430 Direct memory access controller (providing single-channel DMA control)
- 68450 Direct memory access controller (providing four DMA control channels with priorities)
- 68681 Dual channel universal asynchronous receiver/transmitter (UART)

Alternative sources for the 68000 include Hitachi and Signetics. An 8-bit data bus version (68008) is also available. This device is housed in a 48-pin DIL package.

68000 basic configuration

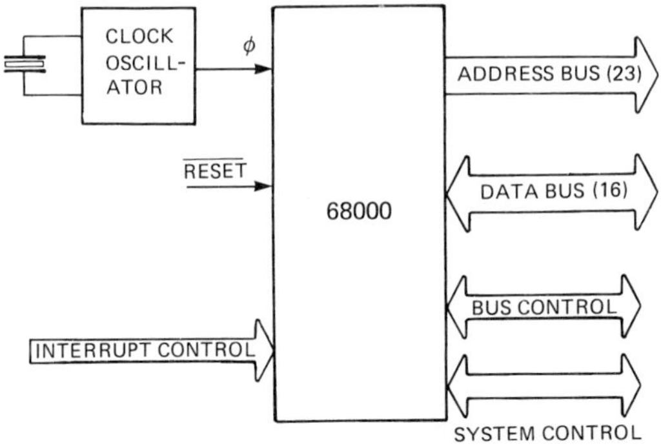

86 Some typical CPUs

68000 register model

DATA REGISTERS

		(D0)
		(D1)
		(D2)
		(D3)
		(D4)
		(D5)
		(D6)
		(D7)

ADDRESS REGISTERS

	(A0)
	(A1)
	(A2)
	(A3)
	(A4)
	(A5)
	(A6)

SPECIAL PURPOSE REGISTERS

USER STACK POINTER	(USP)	(A7)
SUPERVISOR STACK POINTER	(SSP)	(A7)
PROGRAM COUNTER	(PC)	
	STATUS REGISTER	

68000 pinout

```
             ___
D4     ■ 1       64 ■ D5
D3     ■ 2       63 ■ D6
D2     ■ 3       62 ■ D7
D1     ■ 4       61 ■ D8
D0     ■ 5       60 ■ D9
AS     ■ 6       59 ■ D10
UDS    ■ 7       58 ■ D11
LDS    ■ 8       57 ■ D12
R/W    ■ 9       56 ■ D13
DTACK  ■ 10      55 ■ D14
BG     ■ 11      54 ■ D15
BGACK  ■ 12      53 ■ 0 V
BR     ■ 13      52 ■ A23
+5 V   ■ 14      51 ■ A22
Ø      ■ 15      50 ■ A21
0 V    ■ 16      49 ■ +5 V
HALT   ■ 17      48 ■ A20
RESET  ■ 18      47 ■ A19
VMA    ■ 19      46 ■ A18
EN     ■ 20      45 ■ A17
VPA    ■ 21      44 ■ A16
BERR   ■ 22      43 ■ A15
IPL2   ■ 23      42 ■ A14
IPL1   ■ 24      41 ■ A13
IPL0   ■ 25      40 ■ A12
FC2    ■ 26      39 ■ A11
FC1    ■ 27      38 ■ A10
FC0    ■ 28      37 ■ A9
A1     ■ 29      36 ■ A8
A2     ■ 30      35 ■ A7
A3     ■ 31      34 ■ A6
A4     ■ 32      33 ■ A5
```

68000 instruction set

Mnemonic	Description	Operation
ABCD	Add Decimal with Extend	(Destination)$_{10}$ + (Source)$_{10}$ → Destination
ADD	Add Binary	(Destination) + (Source) → Destination
ADDA	Add Address	(Destination) + (Source) → Destination
ADDI	Add Immediate	(Destination) + Immediate Data → Destination
ADDQ	Add Quick	(Destination) + Immediate Data → Destination
ADDX	Add Extended	(Destination) + (Source) + X → Destination
AND	AND Logical	(Destination) ∧ (Source) → Destination
ANDI	AND Immediate	(Destination) ∧ Immediate Data → Destination
ASL, ASR	Arithmetic Shift	(Destination) Shifted by <count> → Destination
B$_{CC}$	Branch Conditionally	If CC then PC + d → PC
BCHG	Test a Bit and Change	~(<bit number>) OF Destination → Z ~(<bit number>) OF Destination → <bit number> OF Destination
BCLR	Test a Bit and Clear	~(<bit number>) OF Destination → Z 0 → <bit number> → OF Destination
BRA	Branch Always	PC + d → PC
BSET	Test a Bit and Set	~(<bit number>) OF Destination → Z 1 → <bit number> OF Destination
BSR	Branch to Subroutine	PC → SP@ −; PC + d → PC
BTST	Test a Bit	~(<bit number>) OF Destination → Z
CHK	Check Register against Bounds	If Dn < 0 or Dn > (<ea>) then TRAP
CLR	Clear an Operand	0 → Destination
CMP	Compare	(Destination) − (Source)
CMPA	Compare Address	(Destination) − (Source)
CMPI	Compare Immediate	(Destination) − Immediate Data
CMPM	Compare Memory	(Destination) − (Source)
DB$_{CC}$	Test Condition, Decrement and Branch	If ~CC then Dn − 1 → Dn; if Dn ≠ −1 then PC + d → PC
DIVS	Signed Divide	(Destination)/(Source) → Destination
DIVU	Unsigned Divide	(Destination)/(Source) → Destination
EOR	Exclusive OR Logical	(Destination) ⊕ (Source) → Destination
EORI	Exclusive OR Immediate	(Destination) ⊕ Immediate Data → Destination
EXG	Exchange Register	Rx ↔ Ry
EXT	Sign Extend	(Destination) Sign-extended → Destination
JMP	Jump	Destination → PC
JSR	Jump to Subroutine	PC → SP@ −; Destination → PC
LEA	Load Effective Address	Destination → An
LINK	Link and Allocate	An → SP@ −; SP → An; SP + d → SP
LSL, LSR	Logical Shift	(Destination) Shifted by <count> → Destination
MOVE	Move Data from Source to Destination	(Source) → Destination
MOVE to CCR	Move to Condition Code	(Source) → CCR
MOVE to SR	Move to the Status Register	(Source) → SR
MOVE from SR	Move from the Status Register	SR → Destination
MOVE USP	Move User Stack Pointer	USP → An, An → USP
MOVEA	Move Address	(Source) → Destination
MOVEM	Move Multiple Registers	Registers → Destination (Source) → Registers
MOVEP	Move Peripheral Data	(Source) → Destination
MOVEQ	Move Quick	Immediate Data → Destination
MULS	Signed Multiply	(Destination)*(Source) → Destination
MULU	Unsigned Multiply	(Destination)*(Source) → Destination
NBCD	Negate Decimal with Extend	0 − (Destination)$_{10}$ − X → Destination
NEG	Negate	0 − (Destination) → Destination
NEGX	Negate with Extend	0 − (Destination) − X → Destination
NOP	No Operation	−
NOT	Logical Complement	~(Destination) → Destination
OR	Inclusive OR Logical	(Destination) ∨ (Source) → Destination
ORI	Inclusive OR Immediate	(Destination) ∨ Immediate Data → Destination
PEA	Push Effective Address	Destination → SP@ −
RESET	Reset External Devices	−
ROL, ROR	Rotate (Without Extend)	(Destination) Rotated by <count> → Destination
ROXL, ROXR	Rotate with Extend	(Destination) Rotated by <count> → Destination
RTE	Return from Exception	SP@ + → SR; SP@ + → PC
RTR	Return and Restore Condition Codes	SP@ + → CC; SP@ + → PC
RTS	Return from Subroutine	SP@ + → PC
SBCD	Subtract Decimal with Extend	(Destination)$_{10}$ − (Source)$_{10}$ − X → Destination
S$_{CC}$	Set According to Condition	If CC then 1's → Destination else 0's → Destination
STOP	Load Status Register and Stop	Immediate Data → SR; STOP
SUB	Subtract Binary	(Destination) − (Source) → Destination
SUBA	Subtract Address	(Destination) − (Source) → Destination
SUBI	Subtract Immediate	(Destination) − Immediate Data → Destination
SUBQ	Subtract Quick	(Destination) − Immediate Data → Destination
SUBX	Subtract with Extend	(Destination) − (Source) − X → Destination
SWAP	Swap Register Halves	Register [31:16] ↔ Register [15:0]
TAS	Test and Set an Operand	(Destination) Tested → CC; 1 → [7] OF Destination
TRAP	Trap	PC → SSP@ −, SR → SSP@ −; (Vector) → PC
TRAPV	Trap on Overflow	If V then TRAP
TST	Test an Operand	(Destination) Tested → CC
UNLK	Unlink	An → SP; SP@ + → An

[ ] = bit number

CPU data

Device coding	Internal architecture	Address bus	Data bus	Function code	Originator/manufacturer	Family type
6500	8			C	Rockwell	6502
6502	8	16	8	G	Mostek	
65F11	8			C	Rockwell	6502
6800	8	16	8	G	Motorola	
6801	8			C	Motorola	6800
6802	8			C	Motorola	6800
6803	8			C	Motorola	6800
6805	8			C	Motorola	6800
6809	8/16	16	8	G	Motorola	6800
8035	8			C	Intel	8080
8039	8			C	Intel	8080
8048	8			C	Intel	8080
8049	8			C	Intel	8080
8080	8	16	8	G	Intel	
8085	8	16	8	G	Intel	8080
8086	16	20	16	G	Intel	(8080)
8088	16	20	8	G	Intel	8086
8096	16			C	Intel	8086
9440	16	16	16	G	Fairchild	
9900	16	16	16	G	Texas	
9980	16	14	8	G	Texas	9900
16008	8	16	8	G	National	(8080)
16016	16	16	16	G	National	(8080)
16032	16	24	16	E	National	
68000	16/32	24	16	E	Motorola	
68008	16/32	20	16	E	Motorola	68000
68010	16/32	24	16	E	Motorola	68000
68020	32	32	32	E	Motorola	(68000)
68200	16/32			C	Motorola	68000
80186	16	20	16	E	Intel	(8086)
80188	16	20	8	E	Intel	(8086)
80286	16	24	16	E	Intel	(8086)
99105	16	16	16	E	Texas	(9900)
CP1600	16	16	16	G	GI	Nova
MN601	16	16	16	G	Data Gen.	Nova
NSC800	8	16	8	E	National	(Z80)
Z8	8			C	Zilog	
Z80	16	16	8	G	Zilog	(8080)
Z8001	16	16	16	G	Zilog	Z8000
Z8002	16	16	16	G	Zilog	Z8000
Z8108	8/16	19	8	E	Zilog	(Z80)
Z8208	8/16	24	8	E	Zilog	(Z80)
Z8116	8/16	19	16	E	Zilog	(Z80)
Z8216	8/16	24	16	E	Zilog	(Z80)

Function codes: C = Microcontroller/microcomputer
E = Enhanced microprocessor
G = General purpose microprocessor

Brackets denote full or partial upward instruction set compatibility.

Support devices

In order to assist in off-loading tasks from the CPU, most modern microcomputer systems employ a number of programmable support devices. These devices fall into the following general classes:

Arithmetic/numeric data co-processors
Arithmetic or numeric data co-processors are capable of performing arithmetic and logical operations on large integer numbers. Floating point operation is also made very much faster due to hardware rather than software implementation.

Clock generators
Specialized clock generators are required by a number of CPUs in order to provide a clock signal with particular characteristics; e.g. some processors require four-phase clocks while others require two-phase clocks in which the individual phases are non-overlapping. This can be much more easily achieved by using a dedicated clock generator chip than by using conventional TTL or CMOS logic.

CRT controllers
A considerable time penalty is paid when the CPU has to stop processing to periodically manipulate the video display. This task can readily be handled by an appropriate support chip which often operates in conjunction with its own reserved area of video RAM.

DMA controllers
Where a sophisticated bus system is employed, and particularly where 'fast' external devices are to be used (e.g. a hard disk), a DMA controller can be used to speed up data transfer by allowing external devices direct access to the system bus.

Graphic/video display processors (GDP/VDP)
These LSI devices are used to produce raster scan displays for video monitors and, in conjunction with a video modulator, television receivers.

The graphic display processor can normally be expected to generate all the necessary video and synchronizing signals as well as managing the storage and retrieval of display data from the dynamic screen-refresh memory.

Enhanced graphic display processors are capable of high speed vector plotting, interfacing with light pens, and organizing a multi-plane (colour, backdrop, pattern, sprite etc.) dynamic memory.

GPIB controllers
Where a microcomputer is to be used with an IEEE-488 General Purpose Instrument Bus, a dedicated LSI controller can be used to handle the necessary protocol and control the GPIB bus/system interface.

Interrupt controllers
On systems employing a range of I/O devices, interrupts can be produced by any one of a number of peripherals. The use of a dedicated interrupt controller can permit faster real-time response and release the CPU from the task of periodically polling all peripheral devices that may require service.

Keyboard controllers
Keyboard controllers perform the mundane but nevertheless

essential task of detecting and managing keyboard input. While some systems consign this task to what may be more properly described as a 'slave' processor, others make use of a specialized keyboard controller.

LAN controllers
LAN controllers perform data link control, data encoding/decoding, and logic-based collision detection associated with the operation of a local area network.

Memory controllers
Memory controllers assist with the management of external memory (of various types). Dynamic RAM controllers may be used to undertake the task of multiplexing the address bus signals (generating the requisite column address strobe, CAS, and row address strobe, RAS, signals) and periodically refreshing the memory.

Programmable parallel interface devices
Programmable parallel interface devices (see separate section) allow parallel (normally byte-wide) transfer of data to/from the system data bus. Devices normally incorporate tri-state bus buffers/drivers and output latches.

Programmable serial interface devices
Programmable serial interface devices (see separate section) facilitate serial data communication between the microcomputer system and remote peripherals or other microcomputers.

Programmable sound generators (PSG)
Given a suitable transducer and interface, a microprocessor is eminently capable of generating sounds. This, however, is a very inefficient solution to the problem of producing audible output from a microcomputer as it needlessly diverts the processor from other more important tasks. Sound generation is therefore a prime example of the need for programmable support devices.

Most programmable sound generators provide a number of internal registers which control the various attributes of the sounds generated (frequency, amplitude, envelope, etc.). In addition, these devices normally cater for several analogue output channels which may be subsequently mixed together before amplification to a level that will drive a loudspeaker.

Real-time clocks
A real-time clock can be implemented using a counter/timer chip in conjunction with a system clock, but a simpler and much more flexible solution is that of making use of a microprocessor-compatible real-time clock device. These are almost invariably CMOS devices and are designed to be operated in conjunction with battery back-up. Internal registers can be loaded with data corresponding to seconds, minutes, hours, days, weeks, months and even years. Interrupts can be programmed to occur at specified time intervals: e.g. from once every second to once every day.

Timers/counters
These devices perform such tasks as time base generation, event counting, and baud rate generation.

I/O control methods

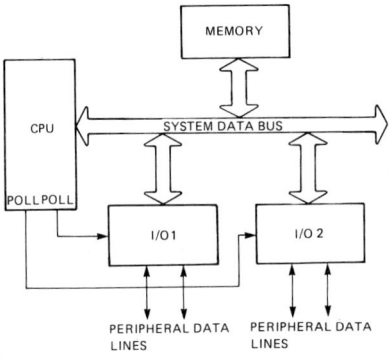

(a) CPU polling

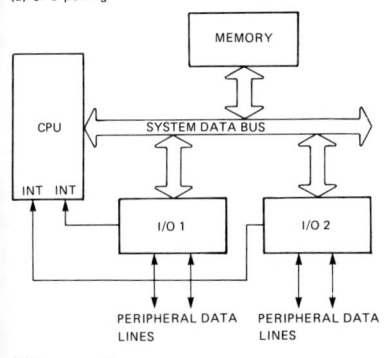

(b) Interrupt driven

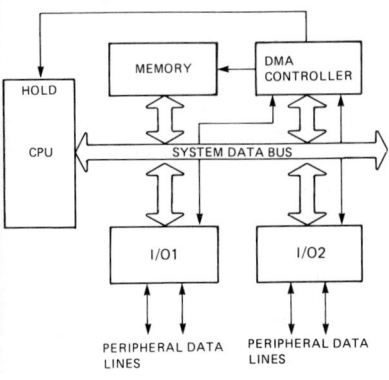

(c) DMA control

Simple parallel I/O interface (Z80-based systems)

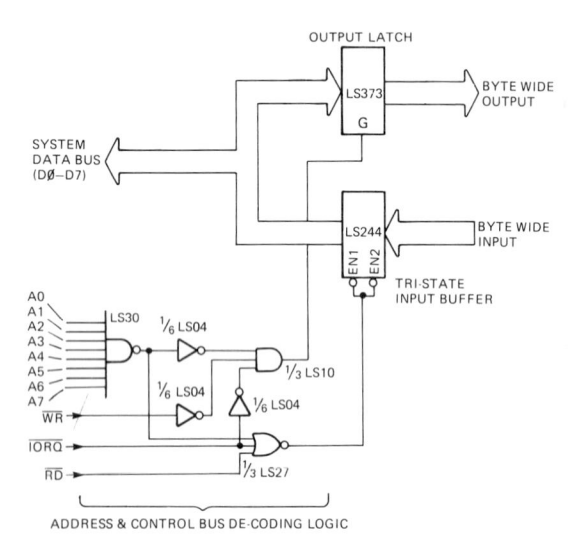

ADDRESS & CONTROL BUS DE-CODING LOGIC

Note: The interface is mapped to a port address of 255. Data is output from the port by first loading the accumulator with the output data and then using an OUT (255),A instruction. Data is input through the port by using an IN A,(255) instruction. The input data then appears in the accumulator.

Programmable parallel interface devices

Most microcomputers incorporate some form of programmable parallel input/output facility. This invariably takes the form of an LSI device variously known as a:

Peripheral interface adaptor	(PIA)	e.g. Motorola 6821
Programmable parallel interface	(PPI)	e.g. Intel 8255
Versatile interface adaptor	(VIA)	e.g. Rockwell 6522
Parallel input/output port	(PIO)	e.g. Zilog Z80-PIO

Such devices generally provide two separate 8-bit ports in which each of the eight bit lines can be configured, under software control, as an input or an output. Additional facilities found in certain more sophisticated devices include automatic handshaking (6522 and Z80-PIO) and internal timers/event counters (6522).

To permit direct connection to the system data bus, the parallel interface device invariably contains a bidirectional tri-state bus interface. Selection of the individual port registers is normally achieved using a subset of the system address bus. The programmable parallel interface thus appears as a number of specific memory addresses which may be selected by appropriate software instructions. The parallel interface device also utilizes the CPU control bus where, for example, a R/W signal is needed in order to determine the direction of data flow from/to the PIA.

The programmable parallel interface device is usually divided internally into two independent sections, A and B. Each section is equipped with three registers, the function of which will be discussed separately. In addition, bidirectional buffers are used to interface the peripheral lines. These buffers are generally TTL compatible and provide a limited current sink of typically 1.6 mA (i.e. one standard TTL load).

During a CPU write operation, the addressed data registers are loaded with the data currently present on the system data bus. The data is then latched onto those lines which have been programmed as outputs. During a CPU read operation the data present on those peripheral lines programmed as inputs is transferred to the system data bus.

The control registers allow the CPU to establish and control the operating modes of the peripheral control lines. In addition, bits are reserved for use as interrupt flags and as a means of selecting either output data or data direction registers. The various bits in the control registers may be accessed many times during a program to allow the CPU to change operating and interrupt modes as required by the particular peripheral device being controlled.

The data direction registers are used to determine which of the peripheral lines are configured as inputs and which are configured as outputs. Each bit position of the data registers corresponds to a particular peripheral line. A logic 1 written to the particular bit position designates the corresponding line as an output, and vice versa. Data direction and data registers often share the same address and selection between the two is made using one of the bits contained in the control register.

Internal registers of a typical programmable parallel I/O device

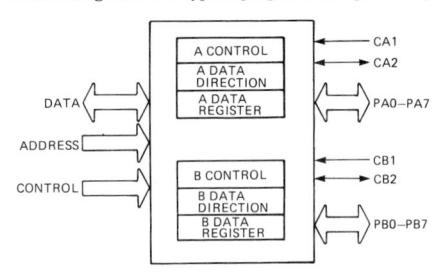

CPU interface to a programmable parallel I/O device

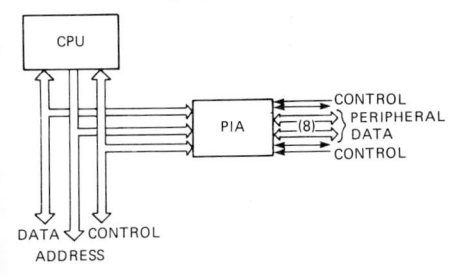

94 Programmable parallel interface devices

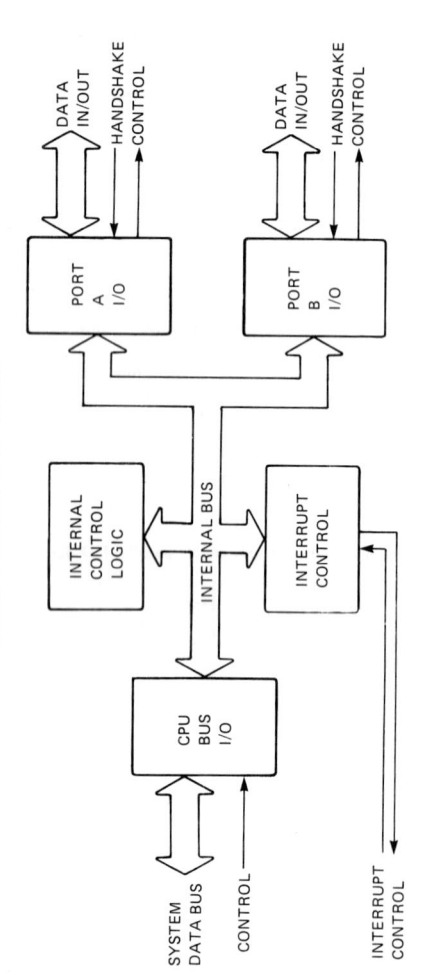

Internal architecture of the Z80-PIO

Programmable serial interface devices 95

Pin connection data for popular programmable parallel I/O devices

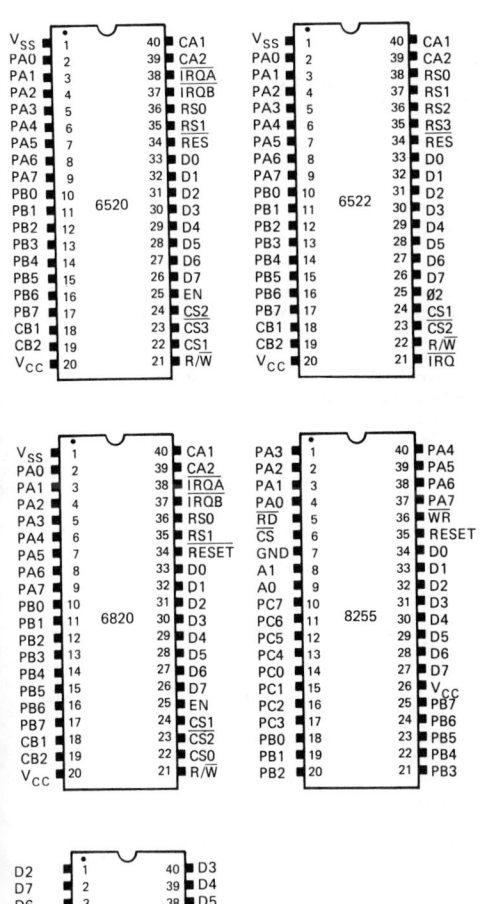

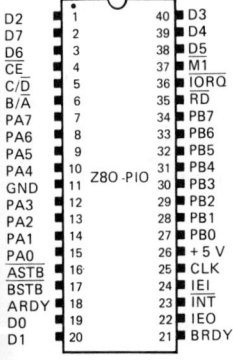

6520

Pin	Signal	Pin	Signal
1	V_SS	40	CA1
2	PA0	39	CA2
3	PA1	38	IRQA
4	PA2	37	IRQB
5	PA3	36	RS0
6	PA4	35	RS1
7	PA5	34	RES
8	PA6	33	D0
9	PA7	32	D1
10	PB0	31	D2
11	PB1	30	D3
12	PB2	29	D4
13	PB3	28	D5
14	PB4	27	D6
15	PB5	26	D7
16	PB6	25	EN
17	PB7	24	CS2
18	CB1	23	CS3
19	CB2	22	CS1
20	V_CC	21	R/W

6522

Pin	Signal	Pin	Signal
1	V_SS	40	CA1
2	PA0	39	CA2
3	PA1	38	RS0
4	PA2	37	RS1
5	PA3	36	RS2
6	PA4	35	RS3
7	PA5	34	RES
8	PA6	33	D0
9	PA7	32	D1
10	PB0	31	D2
11	PB1	30	D3
12	PB2	29	D4
13	PB3	28	D5
14	PB4	27	D6
15	PB5	26	D7
16	PB6	25	Ø2
17	PB7	24	CS1
18	CB1	23	CS2
19	CB2	22	R/W
20	V_CC	21	IRQ

6820

Pin	Signal	Pin	Signal
1	V_SS	40	CA1
2	PA0	39	CA2
3	PA1	38	IRQA
4	PA2	37	IRQB
5	PA3	36	RS0
6	PA4	35	RS1
7	PA5	34	RESET
8	PA6	33	D0
9	PA7	32	D1
10	PB0	31	D2
11	PB1	30	D3
12	PB2	29	D4
13	PB3	28	D5
14	PB4	27	D6
15	PB5	26	D7
16	PB6	25	EN
17	PB7	24	CS1
18	CB1	23	CS2
19	CB2	22	CS0
20	V_CC	21	R/W

8255

Pin	Signal	Pin	Signal
1	PA3	40	PA4
2	PA2	39	PA5
3	PA1	38	PA6
4	PA0	37	PA7
5	RD	36	WR
6	CS	35	RESET
7	GND	34	D0
8	A1	33	D1
9	A0	32	D2
10	PC7	31	D3
11	PC6	30	D4
12	PC5	29	D5
13	PC4	28	D6
14	PC0	27	D7
15	PC1	26	V_CC
16	PC2	25	PB7
17	PC3	24	PB6
18	PB0	23	PB5
19	PB1	22	PB4
20	PB2	21	PB3

Z80-PIO

Pin	Signal	Pin	Signal
1	D2	40	D3
2	D7	39	D4
3	D6	38	D5
4	CE	37	M1
5	C/D	36	IORQ
6	B/A	35	RD
7	PA7	34	PB7
8	PA6	33	PB6
9	PA5	32	PB5
10	PA4	31	PB4
11	GND	30	PB3
12	PA3	29	PB2
13	PA2	28	PB1
14	PA1	27	PB0
15	PA0	26	+5 V
16	ASTB	25	CLK
17	BSTB	24	IEI
18	ARDY	23	INT
19	D0	22	IEO
20	D1	21	BRDY

Programmable serial interface devices

Parallel data transfer is primarily suited to high speed operation over relatively short distances, a typical example being that of linking a microcomputer to an adjacent dot matrix printer. There are, however, a number of applications in which parallel data transfer is inappropriate, the most common example being data communication by means of telephone lines. In such cases data must be sent serially rather than in parallel form. Parallel data from the microprocessor must therefore be reorganized into a train of bits, one after another. An essential requirement of such an arrangement is a means of parallel-to-serial and serial-to-parallel data conversion.

When considering serial transmission of data, a distinction must be made between the two modes of transmission: synchronous and asynchronous. The former method requires a common clock signal to be present at both the sending and receiving ends of the synchronous serial data link. This signal is essential to the decoding process and may either be transmitted along a separate path or regenerated from synchronizing information accompanying the data.

In the asynchronous mode data is sent in a series of small 'packets'. Each packet contains additional 'start' and 'stop' bits which are used to signal the beginning and end of the packet. The position of each bit within the packet may thus be ascertained and the data decoded.

Like their parallel counterparts, programmable serial interface devices generally take the form of a single LSI device known variously as:

Asynchronous communications interface adaptor (ACIA) e.g. Motorola 6850

Synchronous serial data adaptor (SSDA) e.g. Motorola 6852

Universal asynchronous receiver/transmitter (UART) e.g. Intel 8256

Universal synchronous/asynchronous receiver/transmitter (USART) e.g. Intel 8251

Programmable serial interface devices contain a number of registers, including at least one SIPO (serial input, parallel output) and one PISO (parallel input, serial output) shift register. Alternatively, 'universal' shift registers may be employed. These devices can be programmed to operate in either SIPO or PISO modes.

When transmitting data, the appropriate shift register is loaded from the system data bus with data in conventional parallel form. The data is then written out as a serial bit stream by successive shifting.

The reverse process is used for receiving serial data. In this case the incoming data is loaded serially, each successive bit shifting further into the register until it becomes full. Data is then read out simultaneously onto the system bus.

Programmable serial interface devices 97

Parallel to serial data conversion

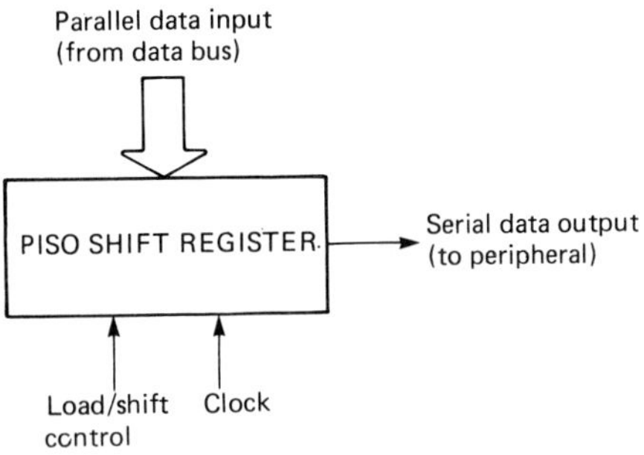

```
Parallel data input
(from data bus)
       ↓
┌──────────────────┐
│ PISO SHIFT REGISTER │ → Serial data output
└──────────────────┘      (to peripheral)
     ↑        ↑
 Load/shift  Clock
  control
```

Serial to parallel data conversion

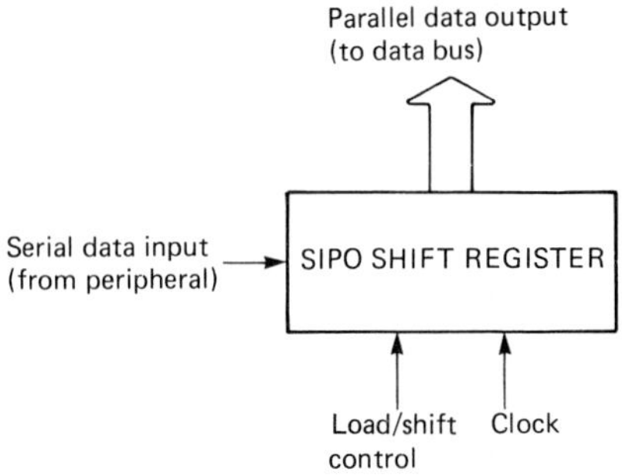

```
              Parallel data output
              (to data bus)
                   ↑
Serial data input  ┌──────────────────┐
(from peripheral) →│ SIPO SHIFT REGISTER │
                   └──────────────────┘
                       ↑        ↑
                   Load/shift  Clock
                    control
```

CPU interface to a programmable serial I/O device

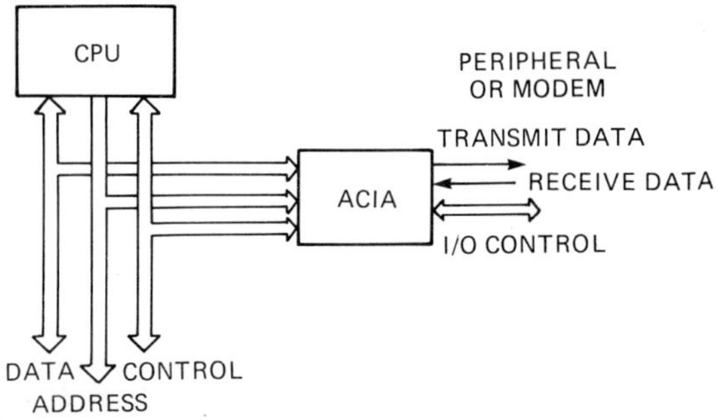

```
┌─────┐                          PERIPHERAL
│ CPU │                          OR MODEM
└─────┘        ┌──────┐   TRANSMIT DATA →
   ⇅⇅⇅    ⇄   │ ACIA │ ← RECEIVE DATA
              └──────┘     I/O CONTROL
   ⇩   ⇩
 DATA  CONTROL
 ADDRESS
```

Pin connection data for popular programmable serial I/O devices

8251

Pin	Signal	Pin	Signal
1	D2	28	D1
2	D3	27	D0
3	RXD	26	V_{CC}
4	GND	25	$\overline{RXC}$
5	D4	24	$\overline{DTR}$
6	D5	23	$\overline{RTS}$
7	D6	22	$\overline{DSR}$
8	D7	21	RESET
9	$\overline{TXC}$	20	CLK
10	$\overline{WR}$	19	TXD
11	$\overline{CS}$	18	TXEMPTY
12	$C/\overline{D}$	17	$\overline{CTS}$
13	$\overline{RD}$	16	SYNDET
14	RXRDY	15	TXRDY

6850

Pin	Signal	Pin	Signal
1	V_{SS}	24	$\overline{CTS}$
2	RXD	23	$\overline{DCD}$
3	RXCLK	22	D0
4	TXCLK	21	D1
5	$\overline{RTS}$	20	D2
6	TXD	19	D3
7	$\overline{IRQ}$	18	D4
8	CS0	17	D5
9	$\overline{CS2}$	16	D6
10	CS1	15	D7
11	RS	14	EN
12	V_{dd}	13	$R/\overline{W}$

Z80–SIO

Pin	Signal	Pin	Signal
1	D1	40	D0
2	D3	39	D2
3	D5	38	D4
4	D7	37	D6
5	$\overline{INT}$	36	$\overline{IORQ}$
6	IEI	35	$\overline{CE}$
7	IEO	34	$B/\overline{A}$
8	$\overline{M1}$	33	$C/\overline{D}$
9	V_{CC}	32	$\overline{RD}$
10	$\overline{W/RDYA}$	31	GND
11	$\overline{SYNCA}$	30	$\overline{W/RDYB}$
12	RXDA	29	RXDB
13	$\overline{RXCA}$	28	$\overline{RXCB}$
14	$\overline{TXCA}$	27	$\overline{TXCB}$
15	TXDA	26	TXDB
16	$\overline{DTRA}$	25	$\overline{DTRB}$
17	$\overline{RTSA}$	24	$\overline{RTSB}$
18	$\overline{CTSA}$	23	$\overline{CTSB}$
19	$\overline{DCDA}$	22	$\overline{DCDB}$
20	CLK	21	$\overline{RESET}$

Basic cell configuration of semiconductor memories

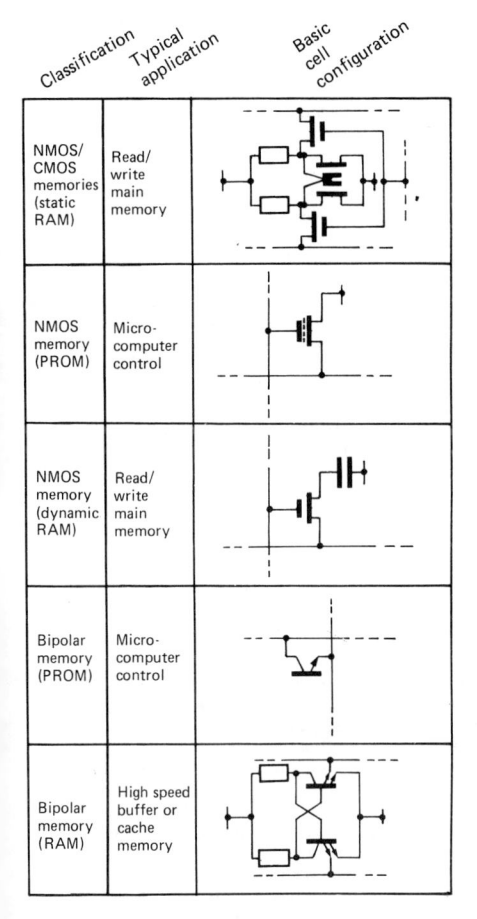

Classification	Typical application	Basic cell configuration
NMOS/CMOS memories (static RAM)	Read/write main memory	
NMOS memory (PROM)	Microcomputer control	
NMOS memory (dynamic RAM)	Read/write main memory	
Bipolar memory (PROM)	Microcomputer control	
Bipolar memory (RAM)	High speed buffer or cache memory	

Semiconductor random access memory

A large proportion (typically 50 per cent or more) of the total addressable memory space of all microcomputers is devoted to read/write memory. This area of memory is used for a variety of purposes, most obvious of which is program and data storage. The term 'random access' simply refers to a memory device in which data may be retrieved from all locations with equal ease (i.e. access time is independent of actual memory address).

The basic element of a semiconductor random access memory is known as a 'cell'. Cells can be fabricated in one of two semiconductor technologies: MOS (metal oxide semiconductor) and bipolar. Bipolar memories are now rarely used in larger memories even though they offer much faster access times. Their disadvantage is associated with power supply requirements, since they need several voltage rails (both positive and negative) and use significantly more power then their MOS counterparts. The high speed of bipolar memories makes them ideal for use in high speed buffer or 'cache' memories. These small capacity memories of typically 4K bytes, or less, permit high speed data transfer without the need for 'wait' states which would be necessary with conventional slower MOS memories.

Random access memories can be further divided into static and dynamic types. The important difference between the two types is that dynamic memories require periodic refreshing if they are not to lose their contents. In the normal course of events, this would be carried out whenever data was read and rewritten, but this technique cannot be relied upon to refresh all of the dynamic memory space and steps must be taken to ensure that all dynamic memory cells are refreshed periodically. This function has to be integrated with the normal operation of the microprocessor. Static memories do not need refreshing and can be relied upon to retain their memory until such time as new data is written or the power supply is interrupted (in which case all data is lost).

Semiconductor RAM data

Type	Size (bits)	Organization	Package	Technology
2102A	1024	1K words × 1 bit	16-pin DIL	S/NMOS
2110	1024	1K words × 1 bit	16-pin DIL	S/ECL
2112	1024	256 words × 4 bits	16-pin DIL	S/ECL
2112A	1024	256 words × 4 bits	16-pin DIL	S/NMOS
2114	4096	1K words × 4 bits	18-pin DIL	S/NMOS
2118	16384	16K words × 1 bit	16-pin DIL	D/NMOS
2128	16384	2K words × 8 bits	24-pin DIL	S/NMOS
2142	4096	4K words × 1 bit	20-pin DIL	S/ECL
2147	4096	4K words × 1 bit	18-pin DIL	S/NMOS
2148	4096	1K words × 4 bits	18-pin DIL	S/NMOS
2149	4096	1K words × 4 bits	18-pin DIL	S/NMOS
2167	16384	16K words × 1 bit	20-pin DIL	S/NMOS
2168	16384	4K words × 4 bits	20-pin DIL	S/NMOS
2169	4096	4K words × 4 bits	20-pin DIL	S/NMOS
2504	256	256 words × 1 bit	16-pin DIL	S/TTL
2510	1024	1K words × 1 bit	16-pin DIL	S/TTL
2511	1024	1K words × 1 bit	16-pin DIL	S/TTL
3764	65536	64K words × 1 bit	16-pin DIL	D/NMOS
4016	16384	2K words × 8 bits	24-pin DIL	S/NMOS
4116	16384	16K words × 1 bit	16-pin DIL	D/NMOS
4118	8192	1K words × 8 bits	24-pin DIL	S/NMOS
4164	65536	64K words × 1 bit	16-pin DIL	D/NMOS

Semiconductor random access memory

Type	Size (bits)	Organization	Package	Technology
4256	262144	256K words × 1 bit	16-pin DIL	D/NMOS
4334	4096	1K words × 4 bits	18-pin DIL	S/CMOS
4364	65536	8K words × 8 bits	28-pin DIL	S/CMOS
4416	65536	16K words × 4 bits	18-pin DIL	D/CMOS
4464	65536	8K words × 8 bits	28-pin DIL	S/CMOS
4716	16384	16K words × 1 bit	16-pin DIL	D/NMOS
4801	8192	1K words × 8 bits	24-pin DIL	S/NMOS
4802	16384	2K words × 8 bits	24-pin DIL	S/NMOS
4816	16384	16K words × 1 bit	16-pin DIL	D/NMOS
4864	65536	64K words × 1 bit	16-pin DIL	D/NMOS
4865	65536	64K words × 1 bit	16-pin DIL	D/NMOS
5101	1024	256 words × 4 bits	22-pin DIL	S/CMOS
5114	4096	1K words × 4 bits	18-pin DIL	S/CMOS
5257	4096	4K words × 1 bit	18-pin DIL	S/CMOS
5516	16384	2K words × 8 bits	24-pin DIL	S/CMOS
5564	65536	8K words × 8 bits	28-pin DIL	S/CMOS
6116	16384	2K words × 8 bits	24-pin DIL	S/CMOS
6117	16384	2K words × 8 bits	24-pin DIL	S/CMOS
6147	4096	4K words × 1 bit	18-pin DIL	S/CMOS
6148	4096	1K words × 4 bits	18-pin DIL	S/CMOS
6167	16384	16K words × 1 bit	20-pin DIL	S/CMOS
6168	16384	4K words × 4 bits	20-pin DIL	S/CMOS
6264	65536	8K words × 8 bits	28-pin DIL	S/CMOS
6267	16384	16K words × 1 bit	20-pin DIL	S/CMOS
6287	65536	64K words × 1 bit	22-pin DIL	S/CMOS
9044	4096	4K words × 1 bit	18-pin DIL	S/NMOS
9064	65536	64K words × 1 bit	16-pin DIL	D/NMOS
9101	1024	256 words × 4 bits	22-pin DIL	S/MOS
9111	1024	256 words × 4 bits	18-pin DIL	S/MOS
9112	1024	256 words × 4 bits	16-pin DIL	S/MOS
9114	4096	1K words × 4 bits	18-pin DIL	S/NMOS
9122	1024	256 words × 4 bits	22-pin DIL	S/MOS
9128	2048	2K words × 8 bits	24-pin DIL	S/MOS
9150	1024	1K words × 4 bits	24-pin DIL	S/NMOS
9151	1024	1K words × 4 bits	24-pin DIL	S/NMOS
99C68	16384	4K words × 4 bits	20-pin DIL	S/CMOS
99C88	65536	8K words × 8 bits	28-pin DIL	S/CMOS
10415	1024	1K words × 1 bit	16-pin DIL	S/ECL
10422	1024	256 words × 4 bits	24-pin DIL	S/ECL
10470	4096	4K words × 1 bit	18-pin DIL	S/ECL
10474	4096	1K words × 4 bits	24-pin DIL	S/ECL
10480	16384	16K words × 1 bit	20-pin DIL	S/ECL
41256	262144	256K words × 1 bit	16-pin DIL	D/NMOS
41257	262144	256K words × 1 bit	16-pin DIL	D/NMOS
41416	65536	16K words × 4 bits	18-pin DIL	D/NMOS
48416	65536	16K words × 4 bits	18-pin DIL	D/NMOS
50256	262144	256K words × 1 bit	16-pin DIL	D/NMOS
50257	262144	256K words × 1 bit	16-pin DIL	D/NMOS
50464	262144	64K words × 4 bits	18-pin DIL	D/NMOS
51258	262144	256K words × 1 bit	18-pin DIL	D/NMOS
65256	262144	32K words × 8 bits	28-pin DIL	S/CMOS
93412	1024	256 words × 4 bits	22-pin DIL	S/TTL
93415	1024	1K words × 1 bit	16-pin DIL	S/TTL

102 Semiconductor random access memory

Pin connection data for popular RAM devices

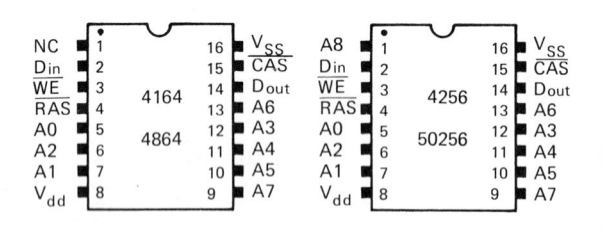

4116 / 4716

Pin	Signal	Pin	Signal
1	V_{bb}	16	V_{SS}
2	D_{in}	15	$\overline{CAS}$
3	$\overline{WE}$	14	D_{out}
4	$\overline{RAS}$	13	A6
5	A0	12	A3
6	A2	11	A4
7	A1	10	A5
8	V_{dd}	9	V_{CC}

4816

Pin	Signal	Pin	Signal
1	NC	16	V_{SS}
2	D_{in}	15	$\overline{CAS}$
3	$\overline{WE}$	14	D_{out}
4	$\overline{RAS}$	13	A6
5	A0	12	A3
6	A2	11	A4
7	A1	10	A5
8	V_{dd}	9	NC

4164 / 4864

Pin	Signal	Pin	Signal
1	NC	16	V_{SS}
2	D_{in}	15	$\overline{CAS}$
3	$\overline{WE}$	14	D_{out}
4	$\overline{RAS}$	13	A6
5	A0	12	A3
6	A2	11	A4
7	A1	10	A5
8	V_{dd}	9	A7

4256 / 50256

Pin	Signal	Pin	Signal
1	A8	16	V_{SS}
2	D_{in}	15	$\overline{CAS}$
3	$\overline{WE}$	14	D_{out}
4	$\overline{RAS}$	13	A6
5	A0	12	A3
6	A2	11	A4
7	A1	10	A5
8	V_{dd}	9	A7

2167

Pin	Signal	Pin	Signal
1	A12	20	V_{CC}
2	A0	19	A13
3	A1	18	A6
4	A2	17	A7
5	A3	16	A8
6	A4	15	A9
7	A5	14	A10
8	Q	13	A11
9	$\overline{WE}$	12	D
10	GND	11	$\overline{CE}$

2168 / 2169

Pin	Signal	Pin	Signal
1	A11	20	V_{CC}
2	A6	19	A10
3	A5	18	A7
4	A4	17	A8
5	A3	16	A9
6	A0	15	I/O1
7	A1	14	I/O2
8	A2	13	I/O3
9	$\overline{CE}$	12	I/O4
10	V_{SS}	11	$\overline{WE}$

6116 / 9128

Pin	Signal	Pin	Signal
1	A7	24	V_{CC}
2	A6	23	A8
3	A5	22	A9
4	A4	21	$\overline{WE}$
5	A3	20	$\overline{OE}$
6	A2	19	A10
7	A1	18	$\overline{CE}$
8	A0	17	IO8
9	IO1	16	IO7
10	IO2	15	IO6
11	IO3	14	IO5
12	GND	13	IO4

6264

Pin	Signal	Pin	Signal
1	NC	28	V_{CC}
2	A12	27	$\overline{WE}$
3	A7	26	CS2
4	A6	25	A8
5	A5	24	A9
6	A4	23	A11
7	A3	22	$\overline{OE}$
8	A2	21	A10
9	A1	20	$\overline{CS1}$
10	A0	19	IO8
11	IO1	18	IO7
12	IO2	17	IO6
13	IO3	16	IO5
14	GND	15	IO4

Typical CMOS battery-backed supply arrangement

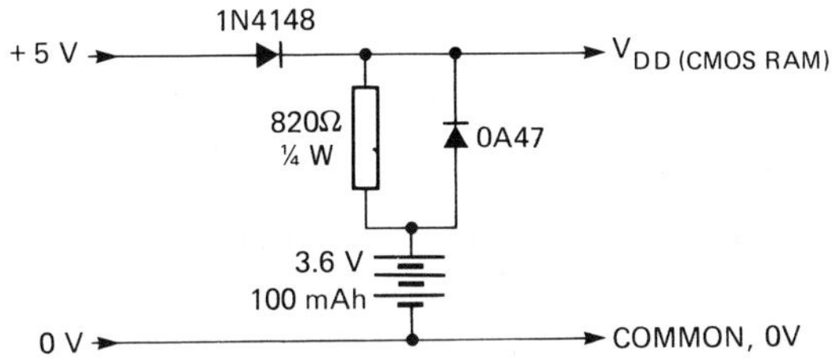

Semiconductor read only memory

As its name implies, read only memory is memory which, once
programmed, can only be read from and not written to. It may
thus be described as 'non-volatile' since its contents are not lost
when the supply is disconnected. This facility is of course necessary
for the long-term semi-permanent storage of operating systems and
high level language interpreters. To change the operating system or
language it is necessary to replace the ROM. This is a simple
matter because ROMs are usually plug-in devices.

The following types are in common use:

Mask ROM

This relatively expensive process is suitable for high-volume
production (several thousand units, or more) and involves the use
of a mask which programs links within the ROM chip. These links
establish a permanent pattern of bits in the row/column matrix of
the memory. The customer (computer manufacturer) must supply
the ROM manufacturer with the programming information from
which the mask is generated.

Programmable ROM (PROM)

This is a somewhat less expensive process than mask programming
and is suitable for small/medium scale production. The memory
cells consist of nichrome or polysilicon fuse links between rows
and columns. These links can, by application of a suitable current
pulse, be open-circuited or 'blown'. PROMs are ideal for prototype
use and programming can be carried out by the computer
manufacturer using relatively inexpensive equipment. When a
PROM has been thoroughly tested, and provided that volume
production can be envisaged, it is normal for the device to be
replaced by a conventional mask programmed ROM.

Erasable PROM (EPROM)

Unlike the two previous types of ROM, the EPROM can be re-
programmed. EPROMs are manufactured with a window which
allows light to fall upon the semiconductor memory cell matrix.
The EPROM may be erased by exposure to a strong ultraviolet
light source over a period of several minutes, or tens of minutes.
Once erasure has taken place, any previously applied bit pattern is
completely removed; the EPROM is 'blank' and ready for
programming. The programming process is carried out by the
manufacturer from master software using a dedicated
programming device which supplies pulses to establish the state of

individual memory cells. This process usually takes several minutes (though some EPROM programmers can program several devices at once) and, since EPROMs tend to be relatively expensive, this process is clearly unsuitable for anything other than very small scale production. Furthermore, it should be noted that EPROMs tend to have rather different characteristics from PROMs and ROMs and thus subsequent volume production replacement may cause problems.

Electrically alterable ROM (EAROM) or electrically erasable programmable ROM (EEPROM)

The EAROM can be both read to and written from. Unlike the random access memory (RAM) the EAROM is unsuitable for use in the read/write memory section of a computer since the writing process takes a considerable time (typically a thousand times longer than the reading time). EAROMs are relatively recent and fairly expensive devices. As such, they have not found many applications in the microcomputer field. It should be noted that a reasonable compromise for semi-permanent data and program storage could take the form of low power consumption CMOS RAM fitted with back-up batteries. In certain circumstances such a system can be relied upon to retain stored information, at relatively low cost, for a year or more. Such an arrangement is an attractive low-cost alternative to the use of EAROM or EEPROM devices.

Semiconductor ROM data

Type	Size (bits)	Organization	Package	Technology
2516	16384	2K words × 8 bits	24-pin DIL	EPROM/NMOS
2532	32768	4K words × 8 bits	24-pin DIL	EPROM/NMOS
2564	65536	8K words × 8 bits	28-pin DIL	EPROM/NMOS
2708	8192	1K words × 8 bits	24-pin DIL	EPROM/NMOS
2716	16384	2K words × 8 bits	24-pin DIL	EPROM/NMOS
27C16	16384	2K words × 8 bits	24-pin DIL	EPROM/CMOS
2732	32768	4K words × 8 bits	24-pin DIL	EPROM/NMOS
27C32	32768	4K words × 8 bits	24-pin DIL	EPROM/NMOS
2764	65536	8K words × 8 bits	28-pin DIL	EPROM/NMOS
27C64	65536	8K words × 8 bits	28-pin DIL	EPROM/CMOS
2816	16384	2K words × 8 bits	24-pin DIL	EEPROM/NMOS
2817	16384	2K words × 8 bits	28-pin DIL	EEPROM/NMOS
27128	131072	16K words × 8 bits	28-pin DIL	EPROM/NMOS
27C128	131072	16K words × 8 bits	28-pin DIL	EPROM/CMOS

Semiconductor read only memory 105

Type	Size (bits)	Organization	Package	Technology
27256	262144	32K words × 8 bits	28-pin DIL	EPROM/NMOS
27C256	262144	32K words × 8 bits	28-pin DIL	EPROM/CMOS
43128	131072	16K words × 8 bits	28-pin DIL	MASK ROM
48016	16384	2K words × 8 bits	24-pin DIL	EAROM/MOS
61366	65536	8K words × 8 bits	24-pin DIL	MASK ROM
613128	131072	16K words × 8 bits	28-pin DIL	MASK ROM
613256	262144	32K words × 8 bits	28-pin DIL	MASK ROM
ER3400	4096	1K words × 4 bits	22-pin DIL	EAROM/MOS

Pin connection data for popular EPROM devices

```
2716                           2732
 A7  1       24  Vcc           A7  1       24  Vcc
 A6  2       23  A8            A6  2       23  A8
 A5  3       22  A9            A5  3       22  A9
 A4  4       21  Vpp           A4  4       21  A11
 A3  5       20  OE            A3  5       20  OE/Vpp
 A2  6       19  A10           A2  6       19  A10
 A1  7       18  CE/PGM        A1  7       18  CE/PGM
 A0  8       17  O7            A0  8       17  O7
 O0  9       16  O6            O0  9       16  O6
 O1 10       15  O5            O1 10       15  O5
 O2 11       14  O4            O2 11       14  O4
 GND 12      13  O3            GND 12      13  O3
```

```
2764                           27128
 Vpp  1      28  Vcc           Vpp  1      28  Vcc
 A12  2      27  PGM           A12  2      27  PGM
 A7   3      26  NC            A7   3      26  A13
 A6   4      25  A8            A6   4      25  A8
 A5   5      24  A9            A5   5      24  A9
 A4   6      23  A11           A4   6      23  A11
 A3   7      22  OE            A3   7      22  OE
 A2   8      21  A10           A2   8      21  A10
 A1   9      20  CE            A1   9      20  CE
 A0  10      19  O7            A0  10      19  O7
 O0  11      18  O6            O0  11      18  O6
 O1  12      17  O5            O1  12      17  O5
 O2  13      16  O4            O2  13      16  O4
 GND 14      15  O3            GND 14      15  O3
```

106 Semiconductor read only memory

```
        27256                              27512
V_PP   1        28  V_CC        A15    1        28  V_CC
A12    2        27  A14         A12    2        27  A14
A7     3        26  A13         A7     3        26  A13
A6     4        25  A8          A6     4        25  A8
A5     5        24  A9          A5     5        24  A9
A4     6        23  A11         A4     6        23  A11
A3     7        22  OE          A3     7        22  OE/V
A2     8        21  A10         A2     8        21  A10
A1     9        20  CE/PGM      A1     9        20  CE/PG
A0    10        19  O7          A0    10        19  O7
O0    11        18  O6          O0    11        18  O6
O1    12        17  O5          O1    12        17  O5
O2    13        16  O4          O2    13        16  O4
GND   14        15  O3          GND   14        15  O3
```

```
            27C1024
V_PP    1           40  V_CC
CE      2           39  PGM
D15     3           38  F
D14     4           37  A15
D13     5           36  A14
D12     6           35  A13
D11     7           34  A12
D10     8           33  A11
D9      9           32  A10
D8     10           31  A9
V_SS   11           30  V_SS
D7     12           29  A8
D6     13           28  A7
D5     14           27  A6
D4     15           26  A5
D3     16           25  A4
D2     17           24  A3
D1     18           23  A2
D0     19           22  A1
OE     20           21  A0
```

Pin connection data for popular EEPROM/EAROM devices

```
        2817A                              9864
R/B    1        28  V_CC        R/B    1        28  V_C
N.C.   2        27  WE          A12    2        27  WE
A7     3        26  N.C.        A7     3        26  N.C
A6     4        25  A8          A6     4        25  A8
A5     5        24  A9          A5     5        24  A9
A4     6        23  N.C.        A4     6        23  A1
A3     7        22  OE          A3     7        22  OE
A2     8        21  A10         A2     8        21  A1
A1     9        20  CE          A1     9        20  CE
A0    10        19  DO7         A0    10        19  DO
DO0   11        18  DO6         DO0   11        18  DC
DO1   12        17  DO5         DO1   12        17  DC
DO2   13        16  DO4         DO2   13        16  DC
GND   14        15  DO3         GND   14        15  DO
```

Storage capacities of mass memories

Medium	Typical capacity (bits)	Typical capacity (bytes)	Equivalent text (A4 pages)	Typical data transfer rate (bits/sec)
Semiconductor ROM	128K	16K	4	4M
Semiconductor RAM	512K	64K	16	4M
Magnetic tape (microdrive)	1M	128K	64	16K
Magnetic tape (0.125 inch cassette)	1M	128K	64	1.2K
Magnetic bubble memory	1M	128K	64	64K
Floppy disk (5.25 ins.)	8M	1M	512	250K
Magnetic tape (0.25 inch cartridge)	16M	2M	1K	100K
Winchester hard disk	80M	10M	5K	4M
Magnetic tape (0.5 inch reel)	256M	32M	16K	800K
WORM laser disk/CD ROM	800M	100M	50K	2.5M

Magnetic disk storage

Magnetic disks are undoubtedly the most popular form of on-line storage for use with computers both large and small. Magnetic disk drives are available in various forms, from the high capacity multi-platter hard disk drives commonly found in mainframe and minicomputer installations to the tiny 3 inch drives used in the smaller personal computers. Despite the obvious outward differences between such devices, essentially they share the same principle of operation; that of writing a magnetic pattern of binary data in the surface oxide coating of a disk. The magnetic pattern is written and read using a small coil contained in a read/write head. The coil itself forms part of a magnetic circuit in which the highest flux density is concentrated in a gap which, depending upon the type of drive, rides either immediately above, or in contact with, the surface of the disk. The main features of each of the popular types of drive are summarized below.

Winchester disks

Today's Winchester disks are the direct descendants of the large multi-platter disk drives used on many mainframe computers of a decade or more ago. Unlike these bulky and extremely expensive units, the modern mini-Winchester drive provides a relatively inexpensive mass storage device in a neat and compact package.

IBM's first Winchester disk drives in the early 1970s were 14 inches in diameter but, in common with the reduction in size of most computer equipment, they rapidly shrank, first to 8 inches and then to 5.25 inches. Storage capacities of up to 160M byte in a package no larger than that occupied by a conventional full-height 5.25 inch floppy disk drive and up to 10M byte in a 3.5 inch micro-Winchester are now commonplace.

The read/write head of a Winchester disk drive rides on a thin cushion of air immediately above the disk surface. This prevents wear on the disk and head surfaces yet maintains efficient magnetic coupling between the head and the disk. To exclude particles of dust, dirt and smoke (which may otherwise lodge between the head and disk surface), the entire disk assembly is housed in a tightly sealed enclosure. Winchester disks rotate at around 3000 rpm and offer fast access times (typically 625K bit/sec).

Unfortunately, Winchester disk drives need rather careful handling since permanent damage can be caused by bumping or jolting the drive (both when in use and when in transit). This is likely to cause the head to 'crash' against the oxide surface of the disk.

Winchester technology has evolved using largely non-exchangeable media, but some effort has been put into producing a removable Winchester cartridge and several manufacturers are working in this area. The problems of maintaining reliability with removable media are, however, not easily overcome!

8 inch floppy disks

The standard 8 inch floppy disk was originally developed by IBM as part of the IBM 3740 key-to-disk data-entry system. These were the forerunners of the ubiquitous 5.25 inch mini-floppy disk. The system uses interchangeable media contained within a low-friction protective mylar sleeve. Storage capacities of between 250K byte and 2M byte are common. In recent years, however, the 8 inch disk has rapidly faded from popularity, having largely been replaced by lower cost 5.25 inch mini-floppy technology.

5.25 inch floppy disks

Mini-floppy disks have a diameter of 5.125 inches (130.2 mm) and the disk sleeve is 5.25 inches (133.4 mm) square. Disks are available in both hard- and soft-sectored formats, the latter enjoying by far the greater popularity.

Hard-sectored disks have a series of index holes which indicate the start of each sector. The sectors on soft-sectored disks have to be written during the formatting process, which effectively writes a framework of tracks and sectors in which the data is subsequently placed.

Mini-floppy disk drives are often categorized by their height in relation to those which first became available (i.e. full-height drives). In recent years there has been a trend towards more compact drives in order to permit a consequent saving of space in the equipment to which they are fitted. This is, of course, particularly important where drives are to be fitted in portable equipment.

The first mini-floppy disk drive to gain popularity was the SA400 from Shugart Associates. This 35 (or 40) track drive provides an unformatted capacity of 125K bytes in single density (FM) or 250K bytes in double density (MFM). The data transfer rates are respectively 125K bit/sec and 250K bit/sec. The drive provides an average latency of 100 ms, a stepping time (track to track) of 20 ms and an average access time of 280 ms. The drive requires d.c. supplies of $+12\,V\ \pm5\%$ at 0.9 A (typical) and $+5\,V\ \pm5\%$ at 0.5 A (typical).

In recent years, the FD-50 series of full-height drives from Teac have achieved immense popularity (the FD-50A being compatible with Shugart's SA400). Half-height drives have also become increasingly popular, with modern devices by Hitachi, Mitsubishi, and Tandon (among others) offering storage capacities well in excess of the SA400 and almost comparable with the smaller hard disks.

3.5 inch floppy disks

The Sony 3.5 inch floppy disk drive was the first 3.5 inch drive to appear in quantity and was largely an extension and improvement of the existing 5.25 inch standard. Many manufacturers have adopted this new standard, which is based upon 40 or 80 tracks at 67.5 or 135 tracks per inch with track–track stepping times of 6 ms and 3 ms respectively. Storage capacities of up to 400K byte per side are thus possible using double density (MFM) recording.

The 3.5 inch disk system is based upon a disk (having less than half the surface area of its 5.25 inch counterpart) housed in a rigid plastic cassette. The disk is provided with an automatic shutter which protects the magnetic surface from accidental contact (a perennial problem with the 5.25 inch mini-floppy disk). Write protection is provided by means of a sliding write-protect tab. The disk is soft-sectored and typically provides approximately 500K byte and 1M byte of unformatted storage on single- and double sided-drives respectively.

3 inch floppy disks

Although the 3.5 inch drive seems to have very largely set the standard for a compact mini-floppy, a number of other drives have appeared at around 3 inches diameter. Most notable among these are Teac's FD30A and Hitachi's HFD30S. These units have both sensibly retained full software formatting compatibility with their 5.25 inch predecessors. Hardware compatibility extends with the use of the standard 34-way disk bus (including PCB edge connector), disk rotation speed and number of tracks per inch.

Other 3 inch drives do exist (including one which uses a single spiral track) but seem unlikely to gain a large following in the industry as a whole.

Microdrives, stringy floppies, and wafadrives

From time to time, attempts have been made to provide a low-cost alternative to disk storage yet one which is faster and more reliable than using compact cassette tapes. Although none of the above methods can be strictly described as disk-based storage media, they all attempt, in some way, to emulate disk storage using a continuous loop of magnetic tape running at high speed.

The first 'stringy floppy' tape drive to be produced in large quantities was that from Exatron, which was designed to be used with the popular Tandy TRS-80 home computer.

None of the few systems currently available can be described as particularly satisfactory (at least when compared with conventional disk storage) and indeed their cost-effectiveness is now somewhat questionable in the light of the falling price of 5.25 inch and 3.5 inch disk drives. Despite this, one major manufacturer (Sinclair) has doggedly promoted its own Microdrive system which is both loved and hated by its devotees!

110 Magnetic disk storage

8 inch disk media format

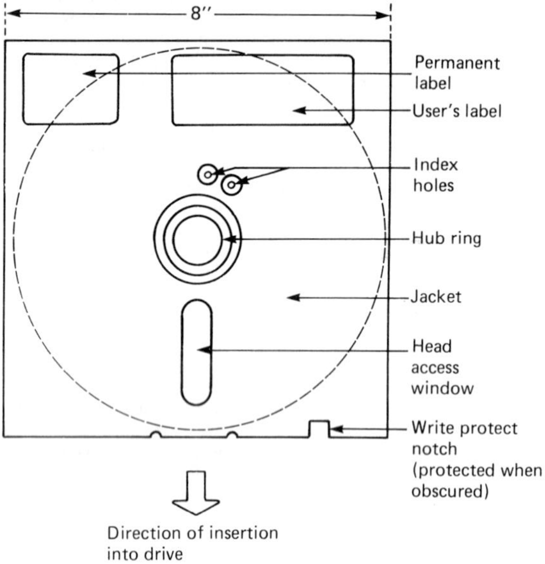

- Permanent label
- User's label
- Index holes
- Hub ring
- Jacket
- Head access window
- Write protect notch (protected when obscured)

Direction of insertion into drive

5.25 inch disk media format

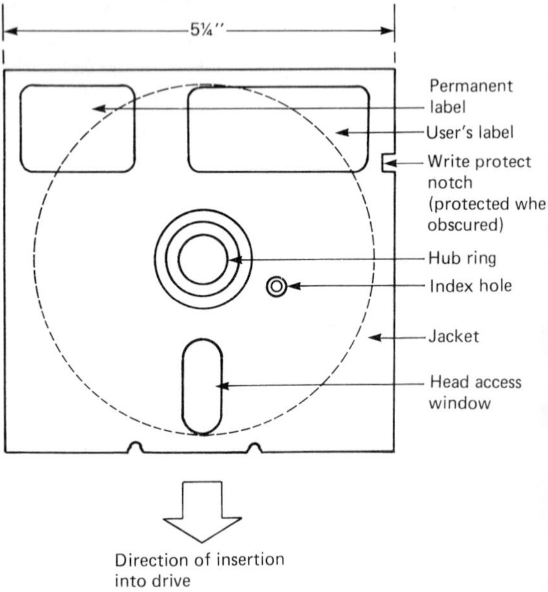

- Permanent label
- User's label
- Write protect notch (protected when obscured)
- Hub ring
- Index hole
- Jacket
- Head access window

Direction of insertion into drive

Magnetic disk storage 111

3.5 inch disk media format

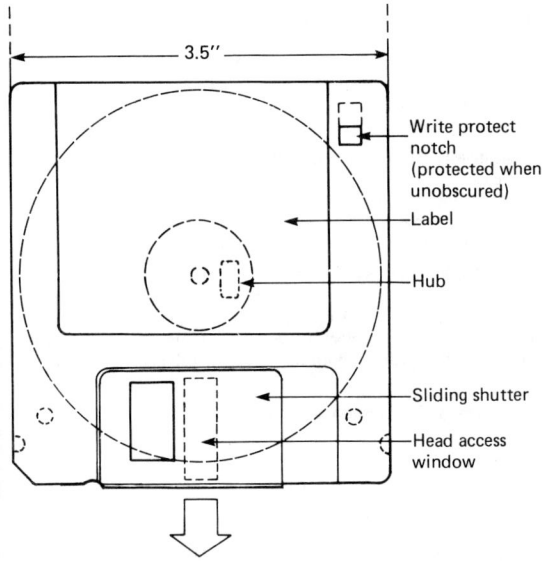

- Write protect notch (protected when unobscured)
- Label
- Hub
- Sliding shutter
- Head access window

Direction of insertion into drive

3 inch disk media format

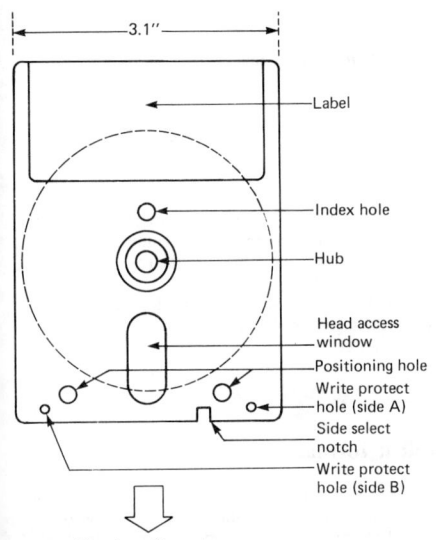

- Label
- Index hole
- Hub
- Head access window
- Positioning hole
- Write protect hole (side A)
- Side select notch
- Write protect hole (side B)

Direction of insertion into drive

Magnetic recording techniques

The two most commonly employed recording techniques use either frequency modulation (FM) or modulated frequency modulation (MFM). Both methods involve writing a serial pulse train (comprising both data and clock pulses) to the disk.

Single density (IBM 3740 standard)
In the case of single-density recording (FM) a clock pulse is present at the start of each bit cell. In this method, a 1 is written by including a pulse in the centre of the cell (i.e. between consecutive clock pulses).

Double density (IBM System 34 standard)
In the double-density recording method things are a little more complex and, whereas a 1 is again written by placing a pulse in the centre of the bit cell, a clock pulse is only written at the start of a cell when a 0 appears in both the preceding cell and the cell in question.

Single-density (FM) recording signal

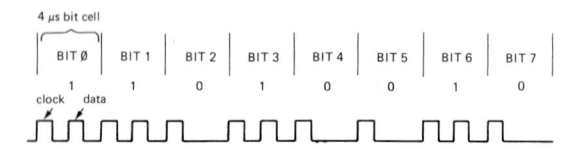

The example shows a byte comprising the hex character D2
Rules: 1. Write a clock bit at the start of each bit cell.
2. If the data is a 1, write a data bit at the centre of the bit cell.

Double-density (MFM) recording signal

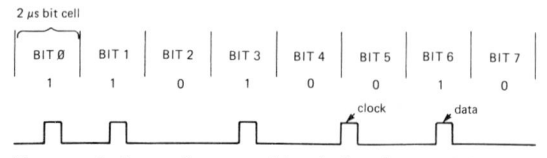

The example shows a byte comprising the hex character D2
Rules: 1. If the data is a 1, write a data bit at the centre of the bit cell.
2. Write a clock bit at the start of a bit cell if no data was written in the last bit cell and no data bit will be written in the next bit cell.

IBM 3740 disk format

The IBM format applies to the vast majority of 8 inch disk systems and, with minor changes (such as the number of tracks and/or sectors), to many of the mini-floppy systems in current use. The main points are listed below.

Track format

The disk has 77 concentric tracks, numbered physically from 00 to 76, with track 00 being the outermost track. During initialization, any two tracks (other than track 00) may be designated as 'bad' and the remaining 75 data tracks are numbered in logical sequence, from 00 to 74.

Sector format

Each track is divided into 26, 15 or 8 sectors of 128, 256, or 512 bytes length respectively. The first sector is numbered 01, and is physically the first sector after the index mark. The remaining sectors are not necessarily numbered in physical sequence, the numbering scheme being determined at initialization.

Each sector consists of a number of fields separated by gaps. An ID field is used to identify the sector, while a data field contains the information stored. The beginning of each field is indicated by six synchronizing bytes of 00H followed by one byte mark.

Address marks

Address marks are unique patterns, one byte in length, which are used to identify the beginning of ID and data fields and to synchronize the de-serializing circuitry with the first byte of each field. Address mark bytes are different from all other data bytes in that certain bit cells do not contain a clock bit (all other data bytes have clock bits in every bit cell). Four different types of address mark are employed to identify different types of field, as follows.

(a) Index address mark The index address mark is located at the beginning of each track and is a fixed number of bytes in advance of the first record. (N.B. Not used in the mini-floppy format.)

(b) ID address mark The ID address mark byte is located at the beginning of each ID field on the disk.

(c) Data address mark The data address mark byte is located at the beginning of each non-deleted data field on the disk.

(d) Deleted data address mark The deleted data address mark byte is located at the beginning of each deleted data field on the disk.

The clock and data patterns used for the various address marks are:

Address mark type	Clock pattern	Data pattern
Index	D7	FC
ID	C7	FE
Data	C7	FB
Deleted data	C7	F8
Bad track ID	C7	FE

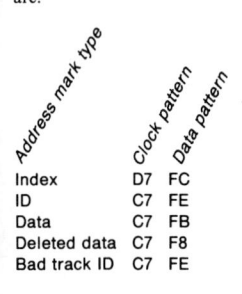

ID field

The ID field precedes the data field and contains a total of seven bytes (including the ID address mark). The ID field contains the track number, side number, sector number, and sector length bytes as well as a two-byte cyclic redundancy code (CRC) checksum. The relationship between the sector length byte and the length of the subsequent data field is:

114 IBM 3740 disk format

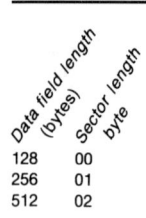

Data field length (bytes)	Sector length byte
128	00
256	01
512	02

Data field
The data field comprises a single-byte data address mark (DAM) followed by the stored data and a trailing two-byte cyclic redundancy code checksum. The data is either 128, 256, or 512 bytes in length.

CRC characters
The 16-bit CRC character is generated using the polynomial $X^{16} + X^{12} + X^5 + 1$, normally initialized to FFH. Its generation includes all characters except the CRC in the ID or data field.

Bad track format
The format is the same as that used for good tracks with the exception that the track number, side number, sector number, and sector length are all set to FFH.

IBM 3740 floppy disk format

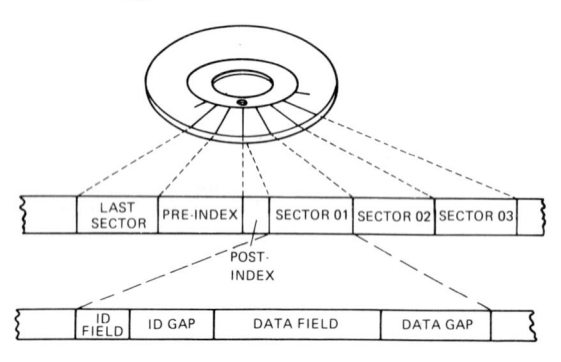

ID field organization (IBM 3740)

One byte

ID ADDRESS MARK	TRACK NUMBER (00–74)	SIDE NUMBER (00–01)	SECTOR NUMBER (01–26)	SECTOR LENGTH (00–02)	CRC (MSB)	CRC (LSB)

Data field organization (IBM 3740)

One byte

DATA ADDRESS MARK	DATA	CRC (MSB)	CRC (LSB)

IBM 3740 format: Gap type and designation

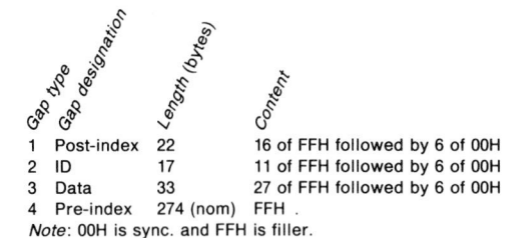

Gap type	Gap designation	Length (bytes)	Content
1	Post-index	22	16 of FFH followed by 6 of 00H
2	ID	17	11 of FFH followed by 6 of 00H
3	Data	33	27 of FFH followed by 6 of 00H
4	Pre-index	274 (nom)	FFH .

Note: 00H is sync. and FFH is filler.

10-sector 80-track single-density format (BBC Microcomputer disk systems)

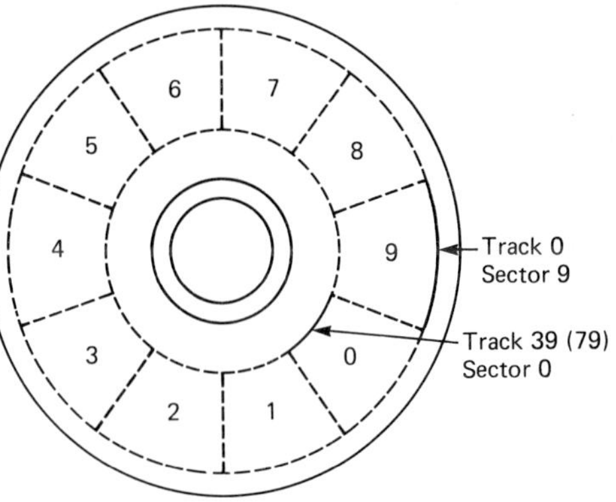

Disk drive mechanics

Floppy disk drives invariably consist of a chassis incorporating the following components:
1. A drive mechanism to rotate the disk at a constant speed.
2. A read/write head mounted on a precision positioning assembly, invariably driven by a stepper motor.
3. Control circuitry which interprets inputs from the disk controller and generates signals to:
(a) start the drive motor
(b) implement write protection so that protected disks cannot be overwritten
(c) drive the head position actuator which moves the head from track to track
(d) activate the head load solenoid which moves the head against the disk
(e) locate the physical index which indicates the start of each track.
4. Read/write circuitry (invariably mounted on the same PCB as the control circuitry) which interfaces the read/write head to produce/accept TTL-compatible signal levels.

A precision servo-controlled d.c. motor is used to maintain the disk speed at 300 rpm ±1.5% for a mini-floppy drive, or 360 rpm ±1.5% for a standard 8 inch floppy drive. Coupling to the disk rotating spindle is normally achieved by means of a rubber drive belt; however, some modern units use direct drive from the motor to the spindle/flywheel assembly.

The drive spindle engages with the centre of the disk (which is usually reinforced by means of additional hub rings) such that the disk rotates within its outer sleeve. The material of the sleeve is chosen so that friction between disk and sleeve is minimal.

Unlike larger hard disks, where the read/write heads ride on a thin cushion of air above the disk surface, the read/write head of a floppy disk makes physical contact with the disk surface (i.e. it is permanently 'crashed') through an elongated hole provided in the sleeve.

The presence (or absence) of a write-protect notch is detected by means of an LED and phototransistor and, if present, write operation is inhibited. Another photo-detector arrangement is used to locate the start of recorded tracks by means of the physical index hole in the disk.

The read/write head assembly is accurately positioned through the use of a precision spiral cam. This cam has a V-groove with a ball-bearing follower which is attached to the head carriage assembly. Precise track selection is accomplished as the cam is rotated in small discrete increments by a stepping motor.

The read/write heads themselves have straddle erase elements which provide erased areas between adjacent data tracks, hence minimizing the effects of data overlap between adjacent tracks when disks written on one drive are read by another.

To ensure a very high degree of compliance with the read/write head, precise registration of the diskette is essential. This is accomplished, with the diskette held in a plane perpendicular to the read/write head, by a platen located in the base casting. The head is loaded against the disk by means of the head load solenoid and a spring-loaded pressure pad is used to maintain contact between the head and the oxide-coated disk surface.

Floppy disk controllers

Floppy disk controllers (FDC) facilitate the storage and retrieval of data in the sectors and tracks, which are written on the disk during the formatting process. The disk controller is thus an extremely complex device, being capable of both formatting the disk and then writing/reading the data on it.

Since neither the user nor the programmer is concerned with the actual organization of data in tracks and sectors, a disk operating system (DOS) is required to carry out management and housekeeping associated with the maintenance of disk files.

The FDC has the following principal functions:

1. Formatting the disk with the required number of tracks and sectors as determined by the DOS.

2. Accepting and executing commands issued by the CPU. These commands are loaded (via the data bus) into a command register within the FDC.

3. Maintaining various internal registers which:

(a) reflect the current status of the controller;

(b) indicate the current track over which the read/write head is positioned; and
(c) hold the address of the desired sector position.

4. Providing an interface to the CPU bus, so that:
(a) during the write process, incoming parallel data from the bus is converted to a serial self-clocking data stream for writing to the floppy disk; and
(b) during the read process, incoming serial data from the floppy disk is separated from the accompanying clock, and fed to a serial-to-parallel shift register before outputting to the data bus.

5. Generating the necessary cyclic redundancy check (CRC) characters and appending these to the write data stream at the appropriate time.

Each disk drive contains its own interface to the bus which links all the drives in a system to the FDC. The most commonly used bus arrangement is that originated by Shugart and first employed with the SA400 drive. This bus uses a 34-way connector, the 17 odd-numbered lines of which are common earth. It should be noted that, since the drive requires an appreciable current, the +12 V and +5 V power lines require a separate connector.

CPU interface to a floppy disk controller

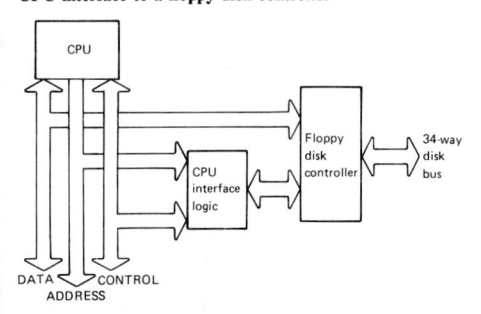

179x floppy disk controller pinout

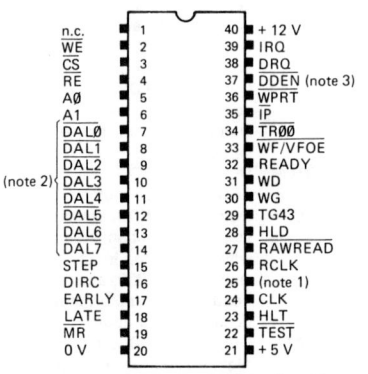

n.c.	1	40	+12 V
$\overline{WE}$	2	39	$\overline{IRQ}$
$\overline{CS}$	3	38	DRQ
$\overline{RE}$	4	37	$\overline{DDEN}$ (note 3)
A0	5	36	$\overline{WPRT}$
A1	6	35	IP
$\overline{DAL0}$	7	34	$\overline{TR00}$
$\overline{DAL1}$	8	33	$\overline{WF}$/VFOE
$\overline{DAL2}$	9	32	READY
(note 2) $\overline{DAL3}$	10	31	WD
$\overline{DAL4}$	11	30	WG
$\overline{DAL5}$	12	29	TG43
$\overline{DAL6}$	13	28	HLD
$\overline{DAL7}$	14	27	RAWREAD
STEP	15	26	RCLK
DIRC	16	25	(note 1)
EARLY	17	24	CLK
LATE	18	23	$\overline{HLT}$
$\overline{MR}$	19	22	$\overline{TEST}$
0 V	20	21	+5 V

Notes: 1. RG when X=1,3; SSO when X=5,7
2. Bus non-inverted when X=3,7
3. Not connected when X=2,4

8271 floppy disk controller pinout

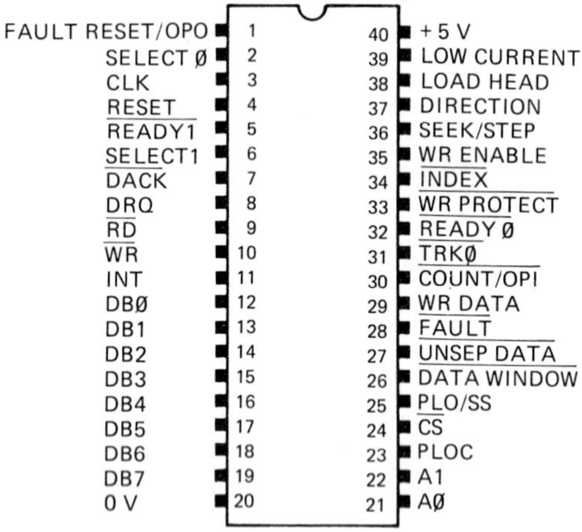

```
FAULT RESET/OPO  1        40  +5 V
      SELECT 0   2        39  LOW CURRENT
           CLK   3        38  LOAD HEAD
         RESET   4        37  DIRECTION
        READY1   5        36  SEEK/STEP
       SELECT1   6        35  WR ENABLE
          DACK   7        34  INDEX
           DRQ   8        33  WR PROTECT
            RD   9        32  READY 0
            WR  10        31  TRK0
           INT  11        30  COUNT/OPI
           DB0  12        29  WR DATA
           DB1  13        28  FAULT
           DB2  14        27  UNSEP DATA
           DB3  15        26  DATA WINDOW
           DB4  16        25  PLO/SS
           DB5  17        24  CS
           DB6  18        23  PLOC
           DB7  19        22  A1
           0 V  20        21  A0
```

34-way Shugart floppy disk bus pin assignment

Pin number	Designation	Common abbrev.	Function
2	Not connected	n.c.	See note 4
4	HEAD LOAD	HLD	Output from FDC, active low, activates the head load solenoid (see note 3)
6	DRIVE SELECT 4	DS4	Output from FDC, active low, selects drive 4
8	INDEX	IP	Input to FDC, active low
10	DRIVE SELECT 1	DS1	Output from FDC, active low, selects drive 1
12	DRIVE SELECT 2	DS2	Output from FDC, active low, selects drive 2
14	DRIVE SELECT 3	DS3	Output from FDC, active low, selects drive 3
16	MOTOR ON	MOTOR	Output from FDC, active low, activates drive motor
18	DIRECTION	DIRC	Output from FDC, selects stepping direction; step out when high, step in when low
20	STEP	STEP	Output from FDC, activates head stepper motor; steps on positive-going edge
22	WRITE DATA	WDATA	Output from FDC, inactive high, pulsed low with data
24	WRITE GATE	WG	Output from FDC, write data when low, read data when high

Floppy disk controllers 119

Pin number	Designation	Common abbrev.	Function
26	TRACK 00	$\overline{\text{TR00}}$	Input to FDC, low when head positioned over track 00
28	WRITE PROTECT	$\overline{\text{WPRT}}$	Input to FDC, active low, indicates that disk has been protected
30	READ DATA	RDATA	Input to FDC, inactive high, pulsed low
32	SIDE SELECT	SIDE	Output from FDC, selects side (double-sided drives only)
34	READY	$\overline{\text{RDY}}$	Input to FDC, active low (see note 3)

Notes: 1. Odd-numbered pins (1 to 33) are 0 V or GROUND.
2. There are some minor variations in the names used for the various lines. In particular, drive selects may be numbered DS0 to DS3 rather than DS1 to DS4.
3. HEAD LOAD and READY signals are not always provided.
4. Pin 2 may either be 'not connected' or 'reserved' for some special purpose.

Pin assignment for a standard 34-way floppy disk bus

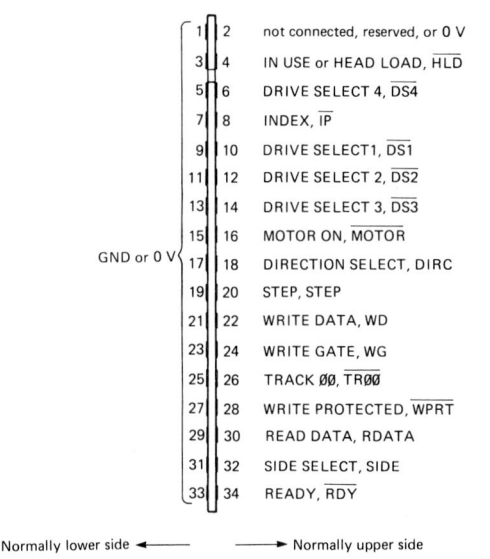

1 | 2	not connected, reserved, or 0 V
3 | 4	IN USE or HEAD LOAD, $\overline{\text{HLD}}$
5 | 6	DRIVE SELECT 4, $\overline{\text{DS4}}$
7 | 8	INDEX, $\overline{\text{IP}}$
9 | 10	DRIVE SELECT 1, $\overline{\text{DS1}}$
11 | 12	DRIVE SELECT 2, $\overline{\text{DS2}}$
13 | 14	DRIVE SELECT 3, $\overline{\text{DS3}}$
15 | 16	MOTOR ON, $\overline{\text{MOTOR}}$
GND or 0 V { 17 | 18	DIRECTION SELECT, DIRC
19 | 20	STEP, STEP
21 | 22	WRITE DATA, WD
23 | 24	WRITE GATE, WG
25 | 26	TRACK 00, $\overline{\text{TR00}}$
27 | 28	WRITE PROTECTED, $\overline{\text{WPRT}}$
29 | 30	READ DATA, RDATA
31 | 32	SIDE SELECT, SIDE
33 | 34	READY, $\overline{\text{RDY}}$

Normally lower side ◄——————► Normally upper side

Note: Edge view of double sided PCB (0.1" pad spacing)

Floppy disk controllers

Disk drive power connections

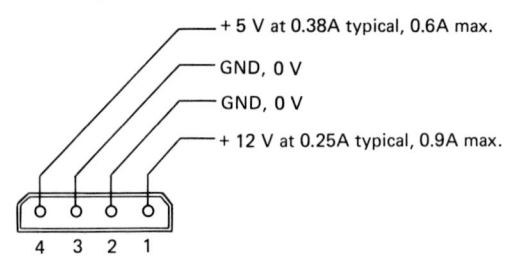

- Pin 1: +12 V at 0.25A typical, 0.9A max.
- Pin 2: GND, 0 V
- Pin 3: GND, 0 V
- Pin 4: +5 V at 0.38A typical, 0.6A max.

Note: Pin view of PCB mounting connector (0.2" pin spacing)

Characteristics of common 5.25 inch floppy disk drives

Drive type	FD-55A compatible FM	FD-55A compatible MFM	FD-55B compatible FM	FD-55B compatible MFM	FD-55E compatible FM	FD-55E compatible MFM	FD-55F compatible MFM
Recording technique	FM	MFM	FM	MFM	FM	MFM	MFM
No. of sides	1	1	2	2	1	1	2
No. of tracks per side	40	40	40	40	80	80	
Track density (tracks per inch)	48	48	48	48	96	96	
Unformatted capacity (bytes) per track	3.125K	6.25K	3.125K	6.25K	3.125K	6.25K	6.25K
per disk	125K	250K	250K	500K	250K	500K	1M
Formatted capacity based on 16 sectors per sector	128	256*	128	256	128	256	256
per track	2K	4K	2K	4K	2K	4K	4K
per disk (bytes)	80K	160K	160K	320K	160K	320K	640K
Data transfer rate (K bit/s)	125	250	125	250	125	250	250
Average access time (ms)	93	93	94	94			
Track access time (ms)	6	6	3	3			
Settling time (ms)	15	15	15	15			

Typical floppy disk capacities

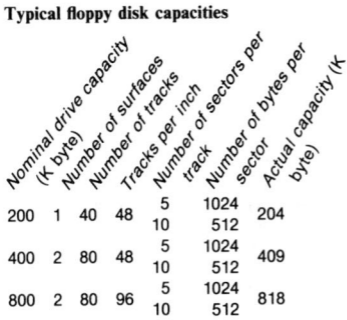

Nominal drive capacity (K byte)	Number of surfaces	Number of tracks	Tracks per inch	Number of sectors per track	Number of bytes per sector	Actual capacity (K byte)
200	1	40	48	5 / 10	1024 / 512	204
400	2	80	48	5 / 10	1024 / 512	409
800	2	80	96	5 / 10	1024 / 512	818

Typical Winchester disk capacities

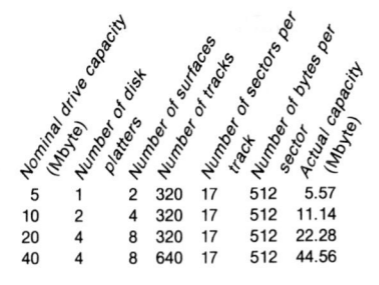

Nominal drive capacity (Mbyte)	Number of disk platters	Number of surfaces	Number of tracks	Number of sectors per track	Number of bytes per sector	Actual capacity (Mbyte)
5	1	2	320	17	512	5.57
10	2	4	320	17	512	11.14
20	4	8	320	17	512	22.28
40	4	8	640	17	512	44.56

Typical pin assignment for a Winchester disk controller

```
49 ────────────//──────────── 1
50 ────────────//──────────── 2
```

Pin number	Signal/function
2	Data line 0
4	Data line 1
6	Data line 2
8	Data line 3
10	Data line 4
12	Data line 5
14	Data line 6
16	Data line 7
36	$\overline{\text{BUSY}}$
38	$\overline{\text{ACKNOWLEDGE}}$
40	$\overline{\text{RESET}}$
42	$\overline{\text{MESSAGE}}$
44	$\overline{\text{SELECT}}$
46	$\overline{\text{COMMAND/DATA}}$
48	$\overline{\text{REQUEST}}$
50	$\overline{\text{INPUT/OUTPUT}}$

Note: All odd-numbered pins (1 to 49) are ground (GND)

Powers of two

n	2^n	n	2^n	n	2^n
0	1	9	512	18	262 144 (256K)
1	2	10	1 024 (1K)	19	524 288 (512K)
2	4	11	2 048 (2K)	20	1 048 576 (1M)
3	8	12	4 096 (4K)	21	2 097 152 (2M)
4	16	13	8 192 (8K)	22	4 194 304 (4M)
5	32	14	16 384 (16K)	23	8 388 608 (8M)
6	64	15	32 768 (32K)	24	16 777 216 (16M)
7	128	16	65 536 (64K)		
8	256	17	131 072 (128K)		

Decimal/hexadecimal/octal/binary/ASCII conversion table

Decimal	Hexadecimal	Octal	Binary	ASCII character
0	00	000	00000000	NUL
1	01	001	00000001	SOH
2	02	002	00000010	STX
3	03	003	00000011	ETX
4	04	004	00000100	EOT
5	05	005	00000101	ENQ
6	06	006	00000110	ACK
7	07	007	00000111	BEL
8	08	010	00001000	BS
9	09	011	00001001	HT
10	0A	012	00001010	LF
11	0B	013	00001011	VT
12	0C	014	00001100	FF
13	0D	015	00001101	CR
14	0E	016	00001110	SO
15	0F	017	00001111	SI
16	10	020	00010000	DLE
17	11	021	00010001	DC1
18	12	022	00010010	DC2
19	13	023	00010011	DC3
20	14	024	00010100	DC4
21	15	025	00010101	NAK
22	16	026	00010110	SYN
23	17	027	00010111	ETB
24	18	030	00011000	CAN
25	19	031	00011001	EM
26	1A	032	00011010	SUB
27	1B	033	00011011	ESC
28	1C	034	00011100	FS
29	1D	035	00011101	GS
30	1E	036	00011110	RS
31	1F	037	00011111	US

Decimal/hexadecimal/octal/binary/ASCII conversion table

Decimal	Hexadecimal	Octal	Binary	ASCII character
32	20	040	00100000	space
33	21	041	00100001	!
34	22	042	00100010	"
35	23	043	00100011	#
36	24	044	00100100	$
37	25	045	00100101	%
38	26	046	00100110	&
39	27	047	00100111	'
40	28	050	00101000	(
41	29	051	00101001	)
42	2A	052	00101010	*
43	2B	053	00101011	+
44	2C	054	00101100	,
45	2D	055	00101101	−
46	2E	056	00101110	.
47	2F	057	00101111	/
48	30	060	00110000	0
49	31	061	00110001	1
50	32	062	00110010	2
51	33	063	00110011	3
52	34	064	00110100	4
53	35	065	00110101	5
54	36	066	00110110	6
55	37	067	00110111	7
56	38	070	00111000	8
57	39	071	00111001	9
58	3A	072	00111010	:
59	3B	073	00111011	;
60	3C	074	00111100	<
61	3D	075	00111101	=
62	3E	076	00111110	>
63	3F	077	00111111	?
64	40	100	01000000	@
65	41	101	01000001	A
66	42	102	01000010	B
67	43	103	01000011	C
68	44	104	01000100	D
69	45	105	01000101	E
70	46	106	01000110	F
71	47	107	01000111	G
72	48	110	01001000	H
73	49	111	01001001	I
74	4A	112	01001010	J
75	4B	113	01001011	K
76	4C	114	01001100	L
77	4D	115	01001101	M
78	4E	116	01001110	N
79	4F	117	01001111	O
80	50	120	01010000	P
81	51	121	01010001	Q
82	52	122	01010010	R
83	53	123	01010011	S
84	54	124	01010100	T

124 Decimal/hexadecimal/octal/binary/ASCII conversion table

85	55	125	01010101	U
86	56	126	01010110	V
87	57	127	01010111	W
88	58	130	01011000	X
89	59	131	01011001	Y
90	5A	132	01011010	Z
91	5B	133	01011011	[
92	5C	134	01011100	\
93	5D	135	01011101	]
94	5E	136	01011110	`
95	5F	137	01011111	_
96	60	140	01100000	'
97	61	141	01100001	a
98	62	142	01100010	b
99	63	143	01100011	c
100	64	144	01100100	d
101	65	145	01100101	e
102	66	146	01100110	f
103	67	147	01100111	g
104	68	150	01101000	h
105	69	151	01101001	i
106	6A	152	01101010	j
107	6B	153	01101011	k
108	6C	154	01101100	l
109	6D	155	01101101	m
110	6E	156	01101110	n
111	6F	157	01101111	o
112	70	160	01110000	p
113	71	161	01110001	q
114	72	162	01110010	r
115	73	163	01110011	s
116	74	164	01110100	t
117	75	165	01110101	u
118	76	166	01110110	v
119	77	167	01110111	w
120	78	170	01111000	x
121	79	171	01111001	y
122	7A	172	01111010	z
123	7B	173	01111011	{
124	7C	174	01111100	|
125	7D	175	01111101	}
126	7E	176	01111110	~
127	7F	177	01111111	DEL
128	80	200	10000000	
129	81	201	10000001	
130	82	202	10000010	
131	83	203	10000011	
132	84	204	10000100	
133	85	205	10000101	
134	86	206	10000110	
135	87	207	10000111	
136	88	210	10001000	
137	89	211	10001001	

138	8A	212	10001010
139	8B	213	10001011
140	8C	214	10001100
141	8D	215	10001101
142	8E	216	10001110
143	8F	217	10001111
144	90	220	10010000
145	91	221	10010001
146	92	222	10010010
147	93	223	10010011
148	94	224	10010100
149	95	225	10010101
150	96	226	10010110
151	97	227	10010111
152	98	230	10011000
153	99	231	10011001
154	9A	232	10011010
155	9B	233	10011011
156	9C	234	10011100
157	9D	235	10011101
158	9E	236	10011110
159	9F	237	10011111
160	A0	240	10100000
161	A1	241	10100001
162	A2	242	10100010
163	A3	243	10100011
164	A4	244	10100100
165	A5	245	10100101
166	A6	246	10100110
167	A7	247	10100111
168	A8	250	10101000
169	A9	251	10101001
170	AA	252	10101010
171	AB	253	10101011
172	AC	254	10101100
173	AD	255	10101101
174	AE	256	10101110
175	AF	257	10101111
176	B0	260	10110000
177	B1	261	10110001
178	B2	262	10110010
179	B3	263	10110011
180	B4	264	10110100
181	B5	265	10110101
182	B6	266	10110110
183	B7	267	10110111
184	B8	270	10111000
185	B9	271	10111001
186	BA	272	10111010
187	BB	273	10111011
188	BC	274	10111100
189	BD	275	10111101
190	BE	276	10111110

Decimal	Hex	Octal	Binary
191	BF	277	10111111
192	C0	300	11000000
193	C1	301	11000001
194	C2	302	11000010
195	C3	303	11000011
196	C4	304	11000100
197	C5	305	11000101
198	C6	306	11000110
199	C7	307	11000111
200	C8	310	11001000
201	C9	311	11001001
202	CA	312	11001010
203	CB	313	11001011
204	CC	314	11001100
205	CD	315	11001101
206	CE	316	11001110
207	CF	317	11001111
208	D0	320	11010000
209	D1	321	11010001
210	D2	322	11010010
211	D3	323	11010011
212	D4	324	11010100
213	D5	325	11010101
214	D6	326	11010110
215	D7	327	11010111
216	D8	330	11011000
217	D9	331	11011001
218	DA	332	11011010
219	DB	333	11011011
220	DC	334	11011100
221	DD	335	11011101
222	DE	336	11011110
223	DF	337	11011111
224	E0	340	11100000
225	E1	341	11100001
226	E2	342	11100010
227	E3	343	11100011
228	E4	344	11100100
229	E5	345	11100101
230	E6	346	11100110
231	E7	347	11100111
232	E8	350	11101000
233	E9	351	11101001
234	EA	352	11101010
235	EB	353	11101011
236	EC	354	11101100
237	ED	355	11101101
238	EE	356	11101110
239	EF	357	11101111
240	F0	360	11110000
241	F1	361	11110001
242	F2	362	11110010
243	F3	363	11110011

244	F4	364	11110100
245	F5	365	11110101
246	F6	366	11110110
247	F7	367	11110111
248	F8	370	11111000
249	F9	371	11111001
250	FA	372	11111010
251	FB	373	11111011
252	FC	374	11111100
253	FD	375	11111101
254	FE	376	11111110
255	FF	377	11111111

ASCII control characters

Hexadecimal	ASCII character	Meaning	Keyboard entry
00	NUL	Null	CTRL-@
01	SOH	Start of heading	CTRL-A
02	STX	Start of text	CTRL-B
03	ETX	End of text	CTRL-C
04	EOT	End of transmission	CTRL-D
05	ENQ	Enquiry	CTRL-E
06	ACK	Acknowledge	CTRL-F
07	BEL	Bell	CTRL-G
08	BS	Backspace	CTRL-H
09	HT	Horizontal tabulation	CTRL-I
0A	LF	Line feed	CTRL-J
0B	VT	Vertical tabulation	CTRL-K
0C	FF	Form feed	CTRL-L
0D	CR	Carriage return	CTRL-M
0E	SO	Shift out	CTRL-N
0F	SI	Shift in	CRTL-O
10	DLE	Data link escape	CTRL-P
11	DC1	Device control one	CTRL-Q
12	DC2	Device control two	CTRL-R
13	DC3	Device control three	CTRL-S
14	DC4	Device control four	CTRL-T
15	NAK	Negative acknowledge	CTRL-U
16	SYN	Synchronous idle	CTRL-V
17	ETB	End of transmission	CTRL-W
18	CAN	Cancel	CTRL-X
19	EM	End of medium	CTRL-Y
1A	SUB	Substitute	CTRL-Z
1B	ESC	Escape	CTRL-[
1C	FS	File separator	CTRL-\
1D	GS	Group separator	CTRL-]
1E	RS	Record separator	CTRL-ˆ
1F	US	Unit separator	CTRL-—

Divisors of 255 with remainders

Divisor (n)	255/n	Remainder	Divisor (n)	255/n	Remainder	Divisor (n)	255/n	Remainder
1	255	0	44	5	35	87	2	81
2	127	1	45	5	30	88	2	79
3	85	0	46	5	25	89	2	77
4	63	3	47	5	20	90	2	75
5	51	0	48	5	15	91	2	73
6	42	3	49	5	10	92	2	71
7	36	3	50	5	5	93	2	69
8	31	7	51	5	0	94	2	67
9	28	3	52	4	47	95	2	65
10	25	5	53	4	43	96	2	63
11	23	2	54	4	39	97	2	61
12	21	3	55	4	35	98	2	59
13	19	8	56	4	31	99	2	57
14	18	3	57	4	27	100	2	55
15	17	0	58	4	23	101	2	53
16	15	15	59	4	19	102	2	51
17	15	0	60	4	15	103	2	49
18	14	3	61	4	11	104	2	47
19	13	8	62	4	7	105	2	45
20	12	15	63	4	3	106	2	43
21	12	3	64	3	63	107	2	41
22	11	13	65	3	60	108	2	39
23	11	2	66	3	57	109	2	37
24	10	15	67	3	54	110	2	35
25	10	5	68	3	51	111	2	33
26	9	21	69	3	48	112	2	31
27	9	12	70	3	45	113	2	29
28	9	3	71	3	42	114	2	27
29	8	23	72	3	39	115	2	25
30	8	15	73	3	36	116	2	23
31	8	7	74	3	33	117	2	21
32	7	31	75	3	30	118	2	19
33	7	24	76	3	27	119	2	17
34	7	17	77	3	24	120	2	15
35	7	10	78	3	21	121	2	13
36	7	3	79	3	18	122	2	11
37	6	33	80	3	15	123	2	9
38	6	27	81	3	12	124	2	7
39	6	21	82	3	9	125	2	5
40	6	15	83	3	6	126	2	3
41	6	9	84	3	3	127	2	1
42	6	3	85	3	0	128	1	127
43	5	40	86	2	83			

Note: This table (and the one that follows) can be used to determine the optimum size for logical records associated with random access disk files. The divisor and remainder give respectively the number of fields (assumed equal size) and the number of wasted bytes.

Divisors of 256 with remainders

Divisor (n)	256/n	Remainder	Divisor (n)	256/n	Remainder	Divisor (n)	256/n	Remainder
1	256	0	44	5	36	87	2	82
2	128	0	45	5	31	88	2	80
3	85	1	46	5	26	89	2	78
4	64	0	47	5	21	90	2	76
5	51	1	48	5	16	91	2	74
6	42	4	49	5	11	92	2	72
7	36	4	50	5	6	93	2	70
8	32	0	51	5	1	94	2	68
9	28	4	52	4	48	95	2	66
10	25	6	53	4	44	96	2	64
11	23	3	54	4	40	97	2	62
12	21	4	55	4	36	98	2	60
13	19	9	56	4	32	99	2	58
14	18	4	57	4	28	100	2	56
15	17	1	58	4	24	101	2	54
16	16	0	59	4	20	102	2	52
17	15	1	60	4	16	103	2	50
18	14	4	61	4	12	104	2	48
19	13	9	62	4	8	105	2	46
20	12	16	63	4	4	106	2	44
21	12	4	64	4	0	107	2	42
22	11	14	65	3	61	108	2	40
23	11	3	66	3	58	109	2	38
24	10	16	67	3	55	110	2	36
25	10	6	68	3	52	111	2	34
26	9	22	69	3	49	112	2	32
27	9	13	70	3	46	113	2	30
28	9	4	71	3	43	114	2	28
29	8	24	72	3	40	115	2	26
30	8	16	73	3	37	116	2	24
31	8	8	74	3	34	117	2	22
32	8	0	75	3	31	118	2	20
33	7	25	76	3	28	119	2	18
34	7	18	77	3	25	120	2	16
35	7	11	78	3	22	121	2	14
36	7	4	79	3	19	122	2	12
37	6	34	80	3	16	123	2	10
38	6	28	81	3	13	124	2	8
39	6	22	82	3	10	125	2	6
40	6	16	83	3	7	126	2	4
41	6	10	84	3	4	127	2	2
42	6	4	85	3	1	128	2	0
43	5	41	86	2	84			

Flowchart symbols

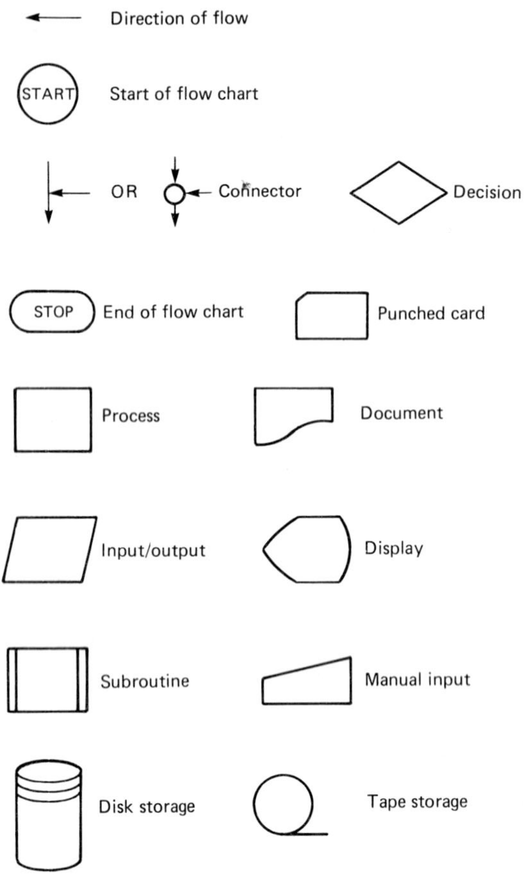

- ← Direction of flow
- (START) Start of flow chart
- ↓← OR ○← Connector
- ◇ Decision
- (STOP) End of flow chart
- Punched card
- Process
- Document
- Input/output
- Display
- Subroutine
- Manual input
- Disk storage
- Tape storage

Operating systems

In general, an operating system is simply a collection of system programs which allow the user to run applications software without having to produce the hardware-specific routines required for such mundane tasks as keyboard and disk I/O. The operating system thus frees the programmer from the need to be aware of the hardware configuration and presents him with what is, in effect, a 'virtual machine' whose characteristics are more tractable than the underlying 'physical machine'.

One obvious advantage of the 'virtual machine' concept is that, provided a form of the operating system is available for a range of machines, a high degree of software portability is ensured.

As a minimum, an operating system can normally be expected to perform the following tasks:

(a) accept keyboard entry of commands and data
(b) create, copy and delete disk files
(c) load programs and data files from disk into RAM
(d) maintain some form of directory of disk files
(e) save programs and data files from RAM to disk.

Operating systems may either be contained in ROM (as 'firmware') or be loaded from disk into RAM when the system is initialized ('booted') on power-up. There are obvious advantages and disadvantages with both of these techniques.

Operating systems normally include a number of commands for manipulating files. These include those for naming, renaming, sorting and copying files. Files are referred to by a 'file specification' (filespec) which, in complete form, comprises:

(a) the drive specification (drivespec) or drive name, comprising a single letter or number;
(b) the filename, which normally comprises a string of up to eight alphanumeric characters;
(c) an extension to the filename, which is normally up to three characters in length.

The extension is usually used to indicate either the type of file or the group of files to which the file belongs. The following are typical examples:

ACC	an accessory file
ASC	a pure ASCII text file
ASM	an assembler source file
BAS	a BASIC program file
BAT	a batch processing file (a file that contains a sequence of system commands which will be automatically executed when the file is loaded)
BIN	a binary file
C	a C program file
CMD	a directly executable utility program (command file)
COM	a compiled object code file/command file
DAT	a data file
DOC	a document file (normally ASCII)
DVR	a driver program
EXE	see CMD
FLT	a device filter
H	a header file (containing predefined symbols and C functions)
IMG	a binary image file for loading directly into memory
LIB	a library file (used in the linking process)
LNK	a link control file
LOG	a LOGO program file
LST	a list file
O	an object code file
OBJ	an object code file
OVn	an overlay (n denotes the overlay number, e.g. OV3 is the third overlay file)
PRG	a directly executable application program (program file)
REL	a relocatable object code file
RND	a random access data file
S	a source code file
SEQ	a sequential access data file
SUB	a submit file (see BAT and CP/M)
SYS	a system file
TXT	a word-processor text file (usually ASCII)

The drivespec is usually followed by a colon or other separator. Where no drivespec is given the drive is taken to be the current (default) drive. The filename is followed by the extension (usually optional). Where the extension is omitted, the system assumes (or supplies) a default extension (CMD when the command line interpreter is active, BAS when the BASIC interpreter is active, and so on). The filename and its extension are normally separated by a full-stop (period) or slash (/).

The following are examples of some complete filenames:

A: CONFIG/SYS	a system file called CONFIG stored in drive A
A: CMDFILE.ASM	an assembler source code file called CMDFILE and contained in drive A
B: INVOICE.BAS	a BASIC program called INVOICE stored in drive B
B: INVOICE.DAT	a data file called INVOICE stored in drive B
HELP.TXT	a text file called HELP stored in the default drive

CP/M

Most microcomputer operating systems can trace their origins in the original CP/M operating system developed by Gary Kildall and written as a software development aid for use with the Intel 8080 microprocessor. In its original form, CP/M was supplied on an 8 inch IBM format floppy disk.

CP/M was subsequently extensively developed and marketed by Digital Research. It now exists in a variety of forms, including those for use with Intel and Motorola 16-bit microprocessor families.

CP/M became the *de facto* operating system for most 8080, 8085 and Z80 8-bit microprocesser-based microcomputers. The wealth of business software written under CP/M guarantees its long term survival, even though it can hardly be described as 'user friendly'.

The following major variants of CP/M are in existence:

CP/M-80	the original 8080-based operating system
CP/M-2.2	improved and enhanced version of CP/M
CP/M-plus	enhanced CP/M with Z80 macro-assembler, faster disk throughput, and banked memory
CP/M-86	CP/M for 8086 family microprocessors
CP/M-68K	CP/M for 68000 family microprocessors
Concurrent CP/M	CP/M which permits time-sharing of the CPU between two or more tasks
MP/M-86	a very much enhanced multi-processing operating system for the 8086 family microprocessors

The following program modules are contained in the CPM.SYS file:

CCP	Console Command Processor (a user interface which parses command lines)
BDOS	Basic Disk Operating System
BIOS	Basic I/O System

CP/M establishes a reserved area of read/write memory which is known as the Transient Program Area (TPA). Application programs are then loaded into this area. The CCP is responsible for communicating with the user, and accepting and acting upon input commands from the console.

CP/M commands

The following represent a typical subset of CP/M commands:

Operating systems 133

A:	specifies drive A as the current disk drive
B:	specifies drive B as the current disk drive
DIR	displays a directory of files contained in the current disk drive (may be appended with a filename or filematch)
DDT	invokes the dynamic debugger (may be appended with a filename)
DUMP	displays the content of a diskfile (the filename must be appended and should include any extension)
ED	invokes the editor (any filename appended must be a text file and any extension must be included)
ERA	erases the specified files (a wild card filematch is provided using an asterisk for either filename or extension; e.g. ERA *.DAT erases all files with the extension DAT from the current disk drive)
FORMAT	formats a disk
LOAD	converts a HEX (the default extension) file into a command file (having a COM extension)
PIP	copies, combines and transfers specified files between peripheral devices
REN	renames an existing file with a specified new name (extensions must be included)
SAVE	saves contents of memory in a disk file (filename and extension must be provided)
STAT	displays status information and/or assigns devices
SUBMIT	a batch processor which executes a sequence of commands contained in the specified submit file (having a SUB extension)
SYSGEN	initializes a system disk
TYPE	displays the contents of a specified ASCII file on the console screen

Notes: 1. Not all of the above commands are appropriate to a particular version of CP/M.
2. Commands such as DIR, ERA, REN, and TYPE are intrinsic CP/M commands.
3. Commands such as FORMAT, PIP, STAT and SUBMIT refer to CP/M utilities.

MS-DOS

With the advent of the PC, IBM opted for a different operating system for use with their 8088/8086-based systems. This operating system is produced by Microsoft and marketed as PC-DOS.

Microsoft later launched MS-DOS for the host of 16-bit PC 'clones' that required a similar operating system to that supplied with the IBM PC. MS-DOS and PC-DOS are thus virtually identical products. The unqualified success of the PC has guaranteed a huge following for MS-DOS, largely at the expense of Digital Research's own CP/M-86.

The following are typical MS-DOS commands:

CAT	sorts, formats and displays a specified disk directory
COPY	copies files from one disk to another (or to the same disk). A wildcard filematch (for either the filename or its extension) is permitted using an asterisk. Various options can be specified, including copying of groups of files; e.g. COPY *.ASC *.TXT copies all files with the extension ASC to new files with the extension TXT

DATE	displays the date
DATE mm-dd-yy	sets a new date
DEL	deletes a file having the appended filename. A drive may also be specified and a wildcard filematch (using an asterisk) is again permitted; e.g. DIR B: *.BAS deletes all files on drive B having a BAS extension
DCOPY	copies an entire disk from one drive to another. Drive specifications must be appended
FORMAT	formats a disk in the specified drive. If -S is appended the newly formatted disk is initialized with the operating system files
REN	renames an old file to a new file; e.g. REN ACC.DAT TAX.DAT renames the old file ACC.DAT to a new file named TAX.DAT. Wildcard renaming is also permitted, using an asterisk in either the filename or extension fields.
TIME	displays the time
TIME hh.mm	sets the time
TYPE	displays the specified file

Note: FORMAT and DCOPY are MS-DOS utilities.

Unix

Unix is a multi-user, multi-tasking operating system which was developed by the Computing Research Group at the Bell Laboratories in New Jersey. Unix was originated on a PDP-11 but is now available for the VAX as well as a number of high performance microcomputers.

Unix has now become synonymous with the C programming language, which has spread from what was essentially a research environment into the wider area of super-microcomputers. In this area it seems set to gain a large following among software developers.

Compared with many of its minicomputer operating system predecessors, Unix is a relatively small operating system. This, however, should not imply that it is in any way lacking in power.

Unix comprises the following main elements:

(a) the 'kernel', which is responsible for management of the resources of the system, including disk drives, printers, terminals, etc.;
(b) the 'file system', which is responsible for file management and the organization of all data storage;
(c) the 'shell' which acts as a bridge between the user and the remainder of the system.

A number of licensed versions of Unix have recently become available, most popular of which is Microsoft's operating system, Xenix.

Software tools

Items of utility software which are available for most microcomputer systems include the following:

Assemblers

An assembler is used to generate machine code (object code) from assembly language source text (source code). The assembler normally needs to read the source text twice in order to accomplish this task. It is then known as a 'two-pass assembler'.

On the first pass, the assembler generates a symbol table which is stored in RAM. This table is used to equate each symbolic address (label) with an absolute address (which is usually not the address at which the program is finally loaded for execution). On the second pass, the assembler generates the machine code (object code) for each instruction.

Various assembler directives can be included within the source code. These pseudo-mnemonics are not translated into object code but are recognized by the assembler during the assembly process. Their purpose is to allow the user to modify, in some way, the object code produced.

The following are some typical Z80 assembler directives:

DEFB exp.	define-byte (the expression or expressions which follow are to be loaded as a byte or bytes into the address held in the location counter; e.g. DEFB 42,64,42)
DEFM string	define-message (loads memory with ASCII values corresponding to a given character string; e.g. DEFM 'Press <ENTER>')
DEFS pppp	define-storage (increases the value held in the location counter by pppp, thus reserving a block of memory of size pppp for storage; e.g. DEFS 100)
DEFW exp.	define-word (as for define-byte but in each case two bytes are loaded and the location counter advances by two)
END	end (marks the end of the code to be assembled)
label EQU exp.	equate (sets the value of the label to be that of the expression; e.g. CR EQU 0DH)
ORG pppp	origin (sets the location counter to pppp and defines the start address in memory at which the following program segment will be resident; e.g. ORG 8000H)

Editors

An editor allows the user to create a text file in a form which may be easily modified. Editors are available in two forms: simple line editors (which only permit operations on a single line of text at a time), and full-screen editors (which permit full cursor control over the whole screen). The action of an editor is similar to that of a simple ASCII word processor. Most assemblers and high level language interpreters will accept pure ASCII text input; thus an editor can be used to create text for programs written in a variety of languages.

Interpreters

An interpreter translates a program written in a high level language (and normally presented in the form of an ASCII text file) into machine executable code. An important feature of an interpreter is that it acts on each statement of source text at a time, i.e. it reads each line, translates it into machine code, and then executes it immediately before fetching the next line of source text. (Where multi-statement lines are permitted each individual statement is interpreted separately.) It should be obvious from this that the interpreter must be used every time the program is run.

The advantage of using an interpreter is that flaws in the source text can be discovered immediately. A statement containing an obvious error will not be executed; the program will simply be halted, an appropriate error message generated, and the user prompted for a correction. The obvious disadvantage of this technique is that, since the interpreter has to act on each line, the program executes comparatively slowly (typically, at one-tenth of the speed when using a compiler). This may, or may not, be significant for a particular application.

Flowchart for an interpreter

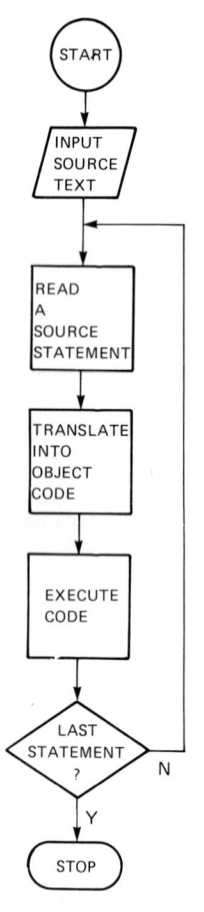

```
START
  │
  ▼
INPUT SOURCE TEXT
  │
  ▼
READ A SOURCE STATEMENT  ◄──┐
  │                          │
  ▼                          │
TRANSLATE INTO OBJECT CODE   │
  │                          │
  ▼                          │
EXECUTE CODE                 │
  │                          │
  ▼                          │
LAST STATEMENT? ──N──────────┘
  │Y
  ▼
STOP
```

Compilers

A compiler can be used to generate a machine code program from source text written in a high level language. The process of compilation is carried out as a separate operation before the program is finally loaded in executable form.

Once the program has been fully debugged, the final compilation process is executed once only. Thereafter, the program is saved, loaded, and run as machine code. Compiled programs therefore run very much faster than those that need the services of an interpreter.

The obvious disadvantage of using a compiler is that, when an error occurs or a change is to be made to the program, the source code must be altered and the entire compilation process (including linking and loading) must be repeated. This tedious process can, to some extent, be shortened with the aid of a batch processing system, in which a single command line can be used to invoke the compiler, linker and loader.

Flowchart for a compiler

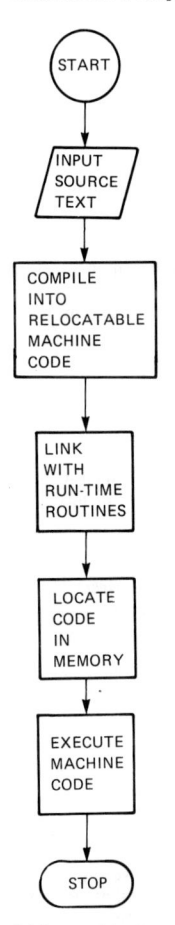

```
START
  ↓
INPUT SOURCE TEXT
  ↓
COMPILE INTO RELOCATABLE MACHINE CODE
  ↓
LINK WITH RUN-TIME ROUTINES
  ↓
LOCATE CODE IN MEMORY
  ↓
EXECUTE MACHINE CODE
  ↓
STOP
```

Linkers and loaders

After compilation, a program usually needs to be linked with the necessary I/O or mathematical routines contained in a 'run-time library'. The necessary routines are then simply added to the machine code. This process is achieved with the aid of a linker (sometimes also called a 'binder').

During compilation and assembly, symbolic labels are normally used to represent addresses. Programs can thus be made relocatable (i.e. they can be loaded anywhere in unreserved RAM). The final stage in the process involves replacing all symbolic addresses with the absolute addresses to be used for final execution. This is achieved with the aid of a loader.

Debuggers

Debuggers provide a means of testing programs interactively during run-time. Common features of debuggers include:
(a) displaying the contents of a given block of memory in hexadecimal or ASCII
(b) loading a program ready for execution
(c) filling a given block of memory with a given data value
(d) commencing execution with optional breakpoints
(e) performing hexadecimal arithmetic
(f) setting up a file control block and command tail
(g) disassembling a block of memory
(h) moving a block of memory from one location to another
(i) reading a disk file into memory
(j) tracing or single-stepping program execution
(k) showing memory layout of a disk file read
(l) writing the contents of a given block of memory to disk
(m) examining and modifying the CPU registers.

A typical 68000-based microcomputer development system would be provided with the following software tools:

AR68.PRG	archive utility (creates library modules that can be linked with LINK68 or LO68)
AS68.PRG	68000 assembler
CP68.PRG CO68.PRG C168.PRG	three-pass C compiler
LINK68.PRG	linker (combines assembled or compiled object modules with appropriate modules taken from the run-time library)
LO68.PRG	another linker
NM68.PRG	symbol table print utility
RELMOD.PRG	loader (produces a program file executable from the GEM operating system)
SID68.PRG	symbolic interactive debugger
SIZE68.PRG	program segment size utility

Note: The .PRG filename extension merely signifies a program file which can be executed directly from the command line interpreter.

Languages

High and low level languages

The choice of programming language required to solve any particular problem depends upon a number of factors, not least of which is the availability of a compiler or interpreter for the machine in question. Happily, a range of languages is available to most modern minicomputers and microcomputers. It is then necessary to make a choice based on such factors as compactness, speed of execution, ease of use, portability, ease of maintenance, etc.

It is fashionable to classify programming languages as either 'low level' or 'high level'. A better distinction would be made between those languages which are 'procedure oriented' and those which are 'machine oriented'.

The lowest level of all programming languages is machine code; i.e. the actual binary coded values that appear in the computer's program memory. The need to code programs in binary is thankfully a thing of the past; indeed, with today's powerful processors it represents a quite impossible task! The nearest approach to using machine code is that of writing programs in assembly language. This task involves using the processor's instruction set presented in mnemonic form.

Unlike assembly language, each statement contained within a program written in a high level language performs some recognizable function. Programs written in a high level language are thus eminently readable. The same cannot be said of assembly language programs unless, of course, they are regularly interspersed with comments.

One statement written in a high level language usually corresponds to many assembly language instructions. The source code for an assembly language program can thus be extremely lengthy. When assembled, however, this code will generally execute many times faster than an interpreted high level language program.

Structured programming

Structured programs are both easier to understand and simpler to maintain than their unstructured counterparts. Most good programming languages contain a variety of logical structures which assist in this respect.

The essential features of structured programs are:

(a) The overall program flow should be sequential. Repeated jumps backwards and forwards within the code should be avoided
(b) Sections of code that are repeatedly executed should be used iteratively; i.e. they should be written out once only and contained within a loop
(c) All transfer of program control should be explicit, using such logical constructs as IF-THEN-ELSE, DO-WHILE, etc.

Assembly language

The most closely related language to that of the machine itself is assembly language. Assembly language programs use symbolic addresses (instead of actual memory locations) and mnemonic operational codes (opcodes). The assembly language program is translated into executable machine code by means of an assembler (see separate section).

The principal disadvantage of assembly language is that programs are not readily transportable from one processor family to another. Furthermore, to be adept with assembly language programs the programmer must have an intimate knowledge of the hardware configuration of a system.

The advantage of assembly language programming is that code is extremely efficient; i.e. it is both fast in execution and very compact. Assembly language programs do not need the services of a compiler or interpreter and thus a minimum of additional software (an assembler and debugger) is required in order to produce a functional program. Indeed, in the case of 8-bit processors, short lengths of code may even be 'hand-assembled'; i.e. the machine code corresponding to a particular assembly language instruction is found by reference to the published instruction set and then simply entered directly into memory using a hexadecimal loader.

Sample assembly language program

```
;   *   SAMPLE ASSEMBLY LANGUAGE PROGRAM   *
;
;   PROGRAM DISPLAYS FULL CHARACTER
;   AND GRAPHIC SET ON THE VDU SCREEN
;
;   @ = SYSTEM CALLABLE ROUTINE
;   $ = POINTER TO STORAGE LOCATION
;
;   Z80 CODE
;
        ORG     0A000H      ;RELOCATABLE
$VIDMEM EQU     3C00H
@DOS    EQU     5500H
        LD      D,9FH
        LD      E,21H
MAIN    LD      HL,$VIDMEM
        LD      BC,400H
LOOP1   LD      (HL),E
        INC     HL
        DEC     BC
        LD      A,B
        OR      C
        JR      NZ,LOOP1
        INC     E
        LD      HL,0FFFFH
LOOP2   DEC     HL
        LD      A,H
        OR      L
        JR      NZ,LOOP2
        DEC     D
        JR      NZ,MAIN
        JP      @DOS
        END
```

APL

APL stands for A Programming Language. For some time APL was used only within IBM and then only on mainframe machines. The reason for this is that the APL interpreter is not particularly compact and, furthermore, it requires a large workspace to run effectively. This fact alone has mitigated against its introduction among the early generations of 8-bit microcomputers. Now, with 16- and 32-bit processors and larger memories, 'micro-APL' is becoming increasingly attractive.

APL was first developed as a consistent notation for mathematics and became available as a programming language in the mid-sixties. APL has some limitations as a language for job processing but does contain an extensive set of operators and data structures. It is thus extremely useful as an investigative tool for use in research and in higher education. The standard APL is VS-APL, although a number of other variants are becoming available. IBM-APL and APL-PLUS are both currently available for use on the IBM PC.

BASIC

BASIC (Beginners All Purpose Symbolic Instruction Code) was developed at Dartmouth College by John Kemeny and Thomas

Kurtz. The principal aim of its creators was to produce a language that the non-programmer would find both acceptable and usable. BASIC statements therefore tend to use simple English words rather than abstract symbols.

BASIC is undoubtedly today's most popular language for home computing and secondary/tertiary education. In its more powerful forms it is capable of handling small business applications with reasonable efficiency.

BASIC exists in an almost infinite number of forms, but the one most widely used is that produced by Microsoft (M-BASIC). Other dialects of BASIC include BBC BASIC, C-BASIC (pseudo-compiled BASIC), Sinclair BASIC and X-BASIC (well suited to multi-user systems).

```
10      REM *** Sample BASIC Program ***
20      ERASE x()
30      INPUT "How many values ";r$
40      LET n=VAL(r$)
50      DIM x(n)
60      FOR i=1 TO n
70      INPUT "Value ";r$
80      LET x(i)=VAL(r$)
90      NEXT i
100     GOSUB 1000
110     GOSUB 2000
120     INPUT "Run again (y/n) ";r$
130     IF r$="n" OR r$="N" THEN END
140     GOTO 10
1000    REM *** Determine Average ***
1010    LET t=0
1020    FOR i=1 TO n
1030    LET t=t+x(i)
1040    NEXT t
1050    PRINT "Average of ";n;" values is ";t/n
1060    RETURN
2000    REM *** Save Array ***
2010    INPUT "Filename for saving ";r$
2020    LET f$=LEFT$(r$,8)
2030    OPEN "O",1,f$
2040    WRITE £1,n
2050    FOR i=1 TO n
2060    WRITE £1,x(i)
2070    NEXT i
2080    CLOSE £1
2090    RETURN
```

C

The C language was developed at the Bell Laboratories in 1972 by Dennis Ritchie. C is based on a language called B, which was itself a development of BCPL (Basic Combined Programming Language).

C uses a relatively small amount of processor-dependent code and is thus highly flexible and portable. The commonly accepted standard for C language programming is the Portable C Compiler (PCC), written by Stephen Johnson. Many versions of Unix use PCC, including the Zilog, Onyx, Xenix, Berkeley Unix and Uniq systems. The Lattice C Compiler is currently the most popular C compiler for use with the IBM PC.

CP/M-68K can run most applications written in C for the Unix operating system, except programs that use the fork/exec multitasking primitives or read Unix file structures.

The accepted reference book for C programming is *The C Programming Language* by Brian Kernighan and Dennis Ritchie (Prentice-Hall).

SAMPLE C PROGRAM

```
/*    SAMPLE C PROGRAM     */

finclude "stdio.h"
finclude "myfile.h"

fdefine QUIT 'Q'

char c;

main()
{
    printf("Press <Q> to quit\n");
    while ((c=getc(stdin))!=QUIT)
    if (isalnum(c))
        decide();
    pause();
    verify();
}

decide()
{
    printf("You pressed %c\n",c);
    if (isdigit (c))
        printf("Its a number!\n");
    else
        printf("Its not a number!\n");
}

verify()
{
    printf("Are you sure (Y/N)?\n");
    while ((c=getc(stdin))!=RETURN)
    switch(c){
    case 'Y':
        exit(0);
    case 'N':
        main();
    default:
        verify;
        break;
    }
}
```

COBOL

COBOL is an acronym for Common Business Oriented Language. COBOL emanated from the Pentagon and is ideally suited to data management. Despite its enormous popularity as an efficient commercial language for use on mainframe and minicomputer installations, COBOL has not made a great impact on the microcomputing world. Despite this, a number of 'micro-COBOLs' have appeared. These include COBOL-80, CIS-COBOL and RM/COBOL.

FORTH

FORTH was invented by Charles Moore as a means of controlling an astronomical telescope at the Kitt Peak Observatory. It is fair to say that no other high level language is as comfortable or versatile in real-time control applications as FORTH. The breadth and scope of its applications are enormous, from controlling a washing machine to managing a canning plant.

A variety of FORTHs exist but most of these conform to the Forth Interest Group 'FIG-FORTH' standard.

FORTH makes extensive use of a resident dictionary and parameter stack. The user is able to define new words and add these to the dictionary. FORTH is thus extensible, and this feature makes it extremely attractive because it allows the user to develop his own application dictionary.

The accepted reference book for the FORTH language is *Starting FORTH* by Leo Brodie (Prentice-Hall).

Sample FORTH program

```
( * SAMPLE FORTH PROGRAM * )
( STEPPER MOTOR DRIVEN BY   )
( Z80 PIO LINES PA0 TO PA2  )
31 CONSTANT DATA
93 CONSTANT CONTROL
: INIT 15 CONTROL OUTP ;
: PUT DATA OUTP ;
: CWSTEP 248 252 PUT PUT ;
: ACWSTEP 250 254 PUT PUT ;
: DELAY 200 0 DO LOOP ;
: CWREV 48 0 DO CWSTEP DELAY LOOP ;
: ACWREV 48 0 DO ACWSTEP DELAY LOOP ;
: CYCLE CWREV ACWREV ;
: PROCESS INIT 10 0 DO CYCLE LOOP ;
```

FORTRAN

FORTRAN is an abbreviation of FORmula TRANslation and has been widely used in science and engineering for more than twenty years. FORTRAN was developed in the 1950s by IBM and since then has been largely restricted to mainframe and minicomputers.

Various versions of FORTRAN are in common use, including FORTRAN IV and, more recently, FORTRAN 77. FORTRAN-80 is available as a microcomputer implementation of the language. This complies with the ANSI standard but does not permit double precision and complex data types. Version 2.00 of the IBM PC FORTRAN compiler conforms to the FORTRAN 77 standard.

LOGO

LOGO was invented by Seymour Papert and reflects the philosophy of the Swiss educationalist, Jean Piaget. LOGO is a graphics-oriented language (famous for its 'turtle') which is eminently suited to providing primary and secondary school children with a meaningful 'hands-on' experience of programming.

It is not surprising, therefore, that LOGO programming has been restricted almost exclusively to the educational world. This, however, belies the power of the language, which bears more than a passing resemblance to FORTH.

Attributes and applications of ten common languages

Attributes

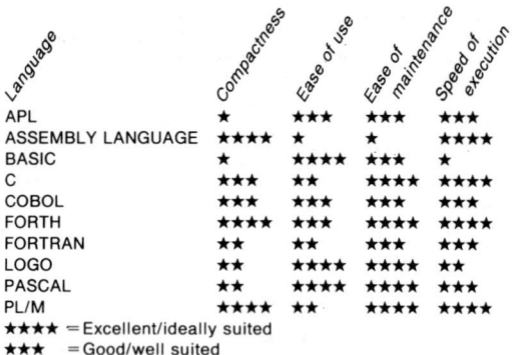

Language	Compactness	Ease of use	Ease of maintenance	Speed of execution
APL	★	★★★	★★★	★★★
ASSEMBLY LANGUAGE	★★★★	★	★	★★★★
BASIC	★	★★★★	★★★	★
C	★★★	★★	★★★★	★★★★
COBOL	★★★	★★★	★★★	★★★
FORTH	★★★★	★★★	★★★★	★★★★
FORTRAN	★★	★★	★★★	★★★
LOGO	★★	★★★★	★★★★	★★
PASCAL	★★	★★★★	★★★★	★★★
PL/M	★★★★	★★	★★★★	★★★★

★★★★ = Excellent/ideally suited
★★★ = Good/well suited
★★ = Poor/dubious choice
★ = Very poor/generally unsuitable

PASCAL

PASCAL was developed by Nicholas Wirth in the early 1970s. He named his language after the French mathematician Blaise Pascal.

The characteristics of PASCAL make it ideal for producing well-structured programs which can be both easily extended and debugged. PASCAL is thus well liked among educationalists and so must now be considered as a worthy contender to BASIC for teaching purposes. PASCAL compiles efficiently and runs several times faster than a comparable program written in FORTRAN or BASIC.

Various versions of PASCAL are currently available for most popular computer systems. Many of these (including PRO-PASCAL) conform to the ANSI/ISO standard with one notable exception, UCSD-PASCAL. This version of the language is highly portable and exhibits considerable machine independence over a wide range of 8- and 16-bit microcomputers.

PL/M

PL/M was devised by Intel in order to facilitate systems software development for its 8-bit microprocessor families. PL/M is itself based on PL/1 (a language which shares some of the features of FORTRAN and COBOL).

PL/M is a block-structured language that is ideally suited to producing modular code. Various PL/M implementations include PL/M-80 (the original 8080-based language) and PL/M-86 (for 8086 family devices). Other manufacturers have produced their own equivalents/alternatives to PL/M, including Motorola (MPL) and Zilog (PLZ).

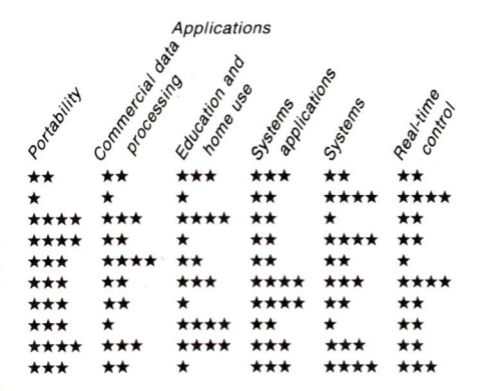

Portability	Commercial data processing	Education and home use	Systems applications	Systems	Real-time control
★★	★★	★★★	★★★	★★	★★
★	★	★	★★	★★★★	★★★★
★★★★	★★★	★★★★	★★	★	★★
★★★★	★★	★	★★	★★★★	★★
★★★	★★★★	★★	★★	★★	★
★★★	★★	★★★	★★★★	★★★	★★★★
★★★	★★	★	★★★★	★★	★★
★★★	★	★★★★	★★	★	★★
★★★★	★★★	★★★★	★★★	★★★	★★
★★★	★★	★	★★★	★★★★	★★★

Video display processing

A typical bit-mapped colour graphics display employs a number of logical memory planes each having a capacity equal to the total number of pixels within the display area.

The display area is given by the number of pixels which appear on the face of the CRT:

Display area = $V \times H$ (pixels)

where V is the number of display area pixels in the vertical direction, and H is the number of display area pixels in the horizontal direction.

Each pixel appearing within the display area is represented by b bits, where b is the number of logical memory planes. Thus, if b planes are provided, the total capacity of reserved video memory is given by:

Total memory required = $V \times h \times n \times b$ (bits)

where h is the number of adjoining n-bit segments corresponding to a line of H pixels ($h = H/n$), and b is the number of logical bit planes.

h memory accesses are required per display line and each access loads b n-bit shift registers.

A simple eight-colour system can be realized using three memory planes (red, green and blue) with appropriate signals clocked out of three shift registers. In more sophisticated systems the shifted output (of four or more planes) is used as an index to a colour 'look-up' palette. The 'looked-up' value is then supplied to three digital to analogue converters which generate the final analogue RGB outputs.

Example 1: A high resolution display based upon the THOMSON-EFCIS EF9365 graphic display processor

$V = H = 512$ pixels
$h = 64$ (i.e. 64 8-bit words per line)

$n = 8$ bits
$b = 3$ (i.e. red, green, and blue memory planes providing 8 colours)

The total video display memory for such a system will be:

$512 \times 64 \times 8 \times 3 = 786\,432$ bits
$= 98\,304$ bytes.

Example 2: Low resolution colour display option employed with the ATARI ST range of microcomputers

$V = 200$ pixels
$H = 320$ pixels
$h = 20$ (i.e. 20 16-bit words per line)
$n = 16$ bits
$b = 4$ (providing 16 colours)

The total video display memory for this system is:

$200 \times 20 \times 16 \times 4 = 256\,000$ bits
$= 32\,000$ bytes.

Logical organization of video memory

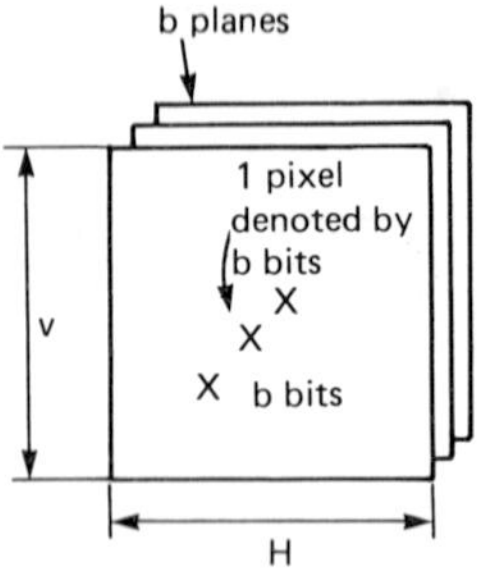

Video display processing (based on EF9365)

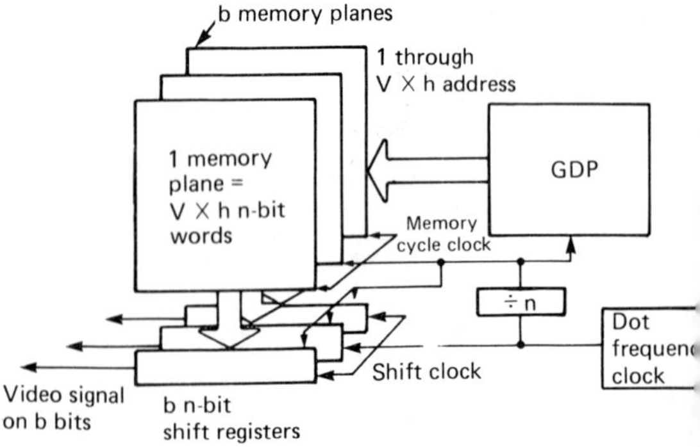

Video resolution

Terms such as 'high', 'medium' and 'standard' are often used to describe the resolution of video monitors. These terms are, however, somewhat misleading and it is advisable to refer to the actual pixel count wherever possible. A further point to note is that the display area produced by a microcomputer is somewhat smaller than the full-face area of the CRT. This imposes a further restriction on the suitability of a particular monitor.

In terms of the total number of pixels ($H \times V$), the following general standards apply:

High resolution   > 450 000 pixels
Medium resolution   300 000 to 450 000 pixels
Low (standard) resolution < 300 000 pixels

Typical high resolution video display

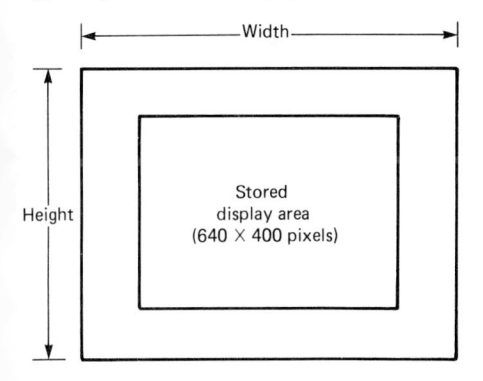

```
Stored
display area
(640 × 400 pixels)
```

(Aspect ratio = Width/Height = 4/3)

Typical video display ASCII character set (based on a 5 × 8 matrix)

Typical video standards

	UK	USA
NON-INTERLACED		
Lines per field	312	260
Lines per frame	312	260
Fields per frame	1	1
Line period	64.1025 μs	64.1025 μs
Line frequency	15.600 kHz	15.600 kHz
Frame period	20 ms	16.666 ms
Frame frequency	50 Hz	60 Hz
Field period	20 ms	16.666 ms
Field frequency	50 Hz	60 Hz
INTERLACED		
Lines per field	312.5	262.5
Lines per frame	625	512
Fields per frame	2	2
Line period	64 μs	63.492 μs
Line frequency	15.625 kHz	15.750 kHz
Frame period	40 ms	33.333 ms
Frame frequency	25 Hz	30 Hz
Field period	20 ms	16.666 ms
Field frequency	50 Hz	60 Hz

Non-interlaced scanning

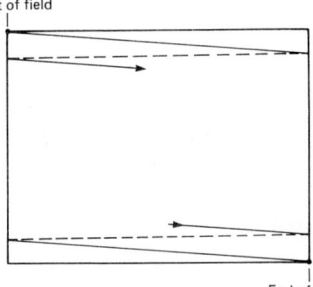

Start of field

End of field

Interlaced scanning

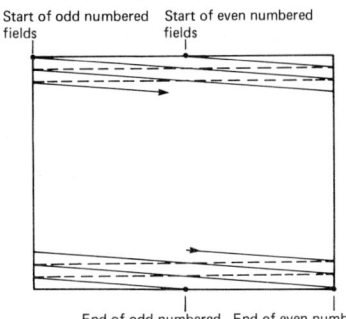

Start of odd numbered fields   Start of even numbered fields

End of odd numbered fields   End of even numbered fields

Typical video waveforms

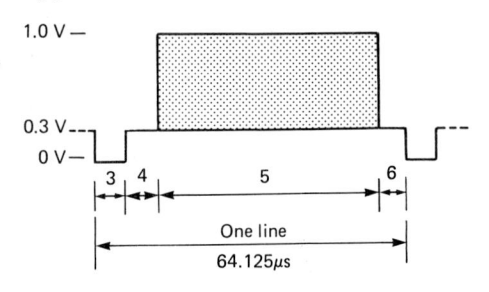

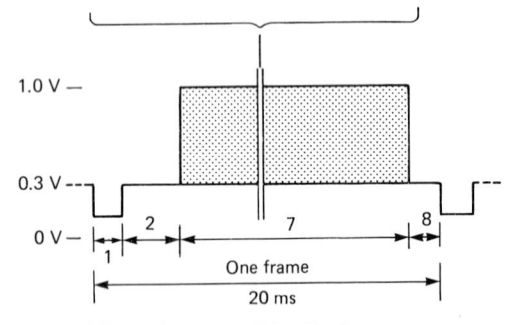

Notes: (a) Frame frequency = 50 Hz, line frequency = 15.625 kHz
  (b) Line sync. pulses not shown on frame waveform
  (c) 1 = frame/vertical sync. pulse
    2 = frame/vertical blanking period
    3 = line/horizontal sync. pulse
    4 = line/horizontal blanking period
    5 = line/horizontal scan/display period
    6 = line/horizontal flyback period
    7 = frame/vertical scan/display period
    8 = frame/vertical flyback period
  (d) Video lines are normally terminated with an impedance of 75 ohms

Typical video waveforms 151

Screen colours and TTL RGB video signals

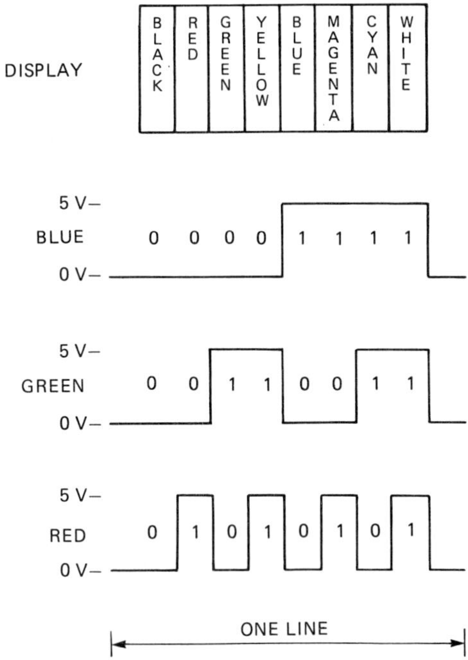

DISPLAY	BLACK	RED	GREEN	YELLOW	BLUE	MAGENTA	CYAN	WHITE

BLUE: 5 V — 0 0 0 0 1 1 1 1 — 0 V

GREEN: 5 V — 0 0 1 1 0 0 1 1 — 0 V

RED: 5 V — 0 1 0 1 0 1 0 1 — 0 V

|←———————— ONE LINE ————————→|

Colour mixing chart

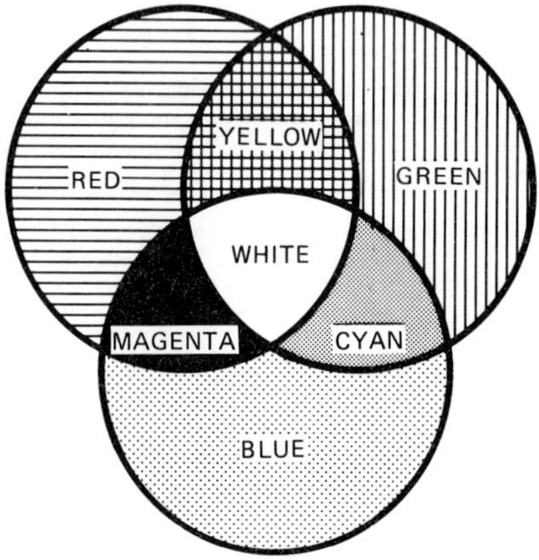

RED, GREEN, BLUE, YELLOW, MAGENTA, CYAN, WHITE

SCART connector pin connections

```
                              21
            Shield ─────────┐ ┌─┐
Composite video output ─── 19 ├─┤ 20 ──── Composite video/sync.
GND (composite video out) ─ 17 ├─┤ 18 ──── GND (fast video blanking)
           Red input ────── 15 ├─┤ 16 ──── Fast video blanking
      GND (red input) ───── 13 ├─┤ 14 ──── GND (data bus)
         Green input ────── 11 ├─┤ 12 ──── Data bus
    GND (green input) ───── 9  ├─┤ 10 ──── Data bus
           Blue input ───── 7  ├─┤ 8  ──── Source switching
      GND (blue input) ──── 5  ├─┤ 6  ──── L. audio input
      L. audio output ───── 3  ├─┤ 4  ──── GND (audio)
      R. audio output ───── 1  ├─┤ 2  ──── R. audio input
```

(Pin view of chassis mounting female connector)

S-100 bus

The S-100 bus was developed by MITS for use in their Altair 8080-based microcomputer system. The system used a number of printed circuit card modules connected into a card frame chassis by means of a double-sided 50-way PCB edge connector. The S-100 bus quickly became the *de facto* standard for the early 8080-based modular microcomputer systems.

The S-100 bus was later refined so that a number of incompatibility problems were reduced (if not eliminated altogether) and the revised standard appears under IEEE-696.

The following points are worthy of note:

(a) The various supply rails are unregulated and thus on-board regulators are required.

(b) There is a risk of cross-talk between some adjacent bus lines on the backplane. For this reason, and to reduce propagation delays in the backplane, the electrical length of the backplane must be reasonably short.

(c) Care must be taken, when inserting and removing cards, to prevent inadvertent shorting of supply rails and signal lines. (In any event this task should *always* be performed with the power off and the reservoir capacitors in a fully discharged state.)

(d) Data lines are unidirectional (rather than bidirectional). Thus two sets of eight data lines ('data in' and 'data out') are provided. If necessary, these two sets of lines can be combined to provide a bidirectional data bus using an 8-bit bidirectional bus transceiver.

(e) The revised S-100 standard (IEEE-696) makes provision for 16-bit processors by extending the address bus from 16 to 24 bits and by ganging the 8-bit data-in and data-out buses into a 16-bit bidirectional bus. Additional handshake lines are included in order to permit intermixing of 8 and 16-bit memory cards.

(f) The data rate of any signal on the bus should not exceed 6 MHz.

(g) The overall structure of the bus is:
  (i) 16 data lines
  (ii) 16 (or 24) address lines
  (iii) 8 status lines
  (iv) 19 control lines (5 output, 6 input and 8 DMA)
  (v) 8 vectored interrupt lines
  (vi) 20 utility bus lines
  (vii) 5 power lines.
(h) Four lines are reserved for future use and three further lines are undefined and are thus available for use by individual manufacturers who should clearly specify their logical function. The voltage level of any signal on these lines should not exceed 5 V.

The S-100 Bus Handbook by Dave Bursky (Hayden) provides a useful reference to the S-100 bus and includes a number of circuit diagrams of representative S-100 card modules.

S-100 pin assignment

Pin No.	Abbreviation	Active level	Signal/function
1	+8 V		Unregulated supply rail
2	+18 V		Unregulated supply rail
3	XRDY	H	Ready input to bus master
4	VI0	L*	Vectored interrupt line 0
5	VI1	L*	Vectored interrupt line 1
6	VI2	L*	Vectored interrupt line 2
7	VI3	L*	Vectored interrupt line 3
8	VI4	L*	Vectored interrupt line 4
9	VI5	L*	Vectored interrupt line 5
10	VI6	L*	Vectored interrupt line 6
11	VI7	L*	Vectored interrupt line 7
12	NMI	L*	Non-maskable interrupt
13	PWRFAIL	L	Power fail signal (pulled low when a power failure is detected)
14	DMA3	L*	DMA request line with highest priority
15	A18		Extended address bus line 18
16	A16		Extended address bus line 16
17	A17		Extended address bus line 17
18	SDSB	L*	Status disable (tri-states all status lines)
19	CDSB	L*	Command disable (tri-states all control input lines)
20	GND		Common 0 V line
21	NDEF		Undefined
22	ADSB	L*	Address disable (tri-states all address lines)
23	DODSB	L*	Data out disable (tri-states all data output lines)
24	∅		Bus clock
25	pSTVAL	L	Status valid strobe (indicates that status information is true)
26	pHLDA	H	Hold acknowledge (signal from the current bus master which indicates that control will pass to the device seeking bus control on the next bus cycle)

Pin No.	Abbreviation	Active level	Signal/function
27	RFU		Reserved for future use
28	RFU		Reserved for future use
29	A5		Address line 5
30	A4		Address line 4
31	A3		Address line 3
32	A15		Address line 15
33	A12		Address line 12
34	A9		Address line 9
35	DO1/Data 1		Data out line 1/bidirectional data line 1
36	DO0/Data 0		Data out line 0/bidirectional data line 0
37	A10		Address line 10
38	DO4/Data 4		Data out line 4/bidirectional data line 4
39	DO5/Data 5		Data out line 5/bidirectional data line 5
40	DO6/Data 6		Data out line 6/bidirectional data line 6
41	DI2/Data 10		Data in line 2/bidirectional data line 10
42	DI3/Data 11		Data in line 3/bidirectional data line 11
43	DI7/Data 15		Data in line 7/bidirectional data line 15
44	sM1	H	M1 cycle (indicates that the current machine cycle is an operation code fetch)
45	sOUT	H	Output (indicates that data is being transferred to an output device)
46	sINP	H	Input (indicates that data is being fetched from an input device)
47	sMEMR	H	Memory read (indicates that the bus master is fetching data from memory)
48	sHLTA	H	Halt acknowledge (indicates that the bus master is executing an HLT instruction)
49	CLOCK		2 MHz clock
50	GND		Common 0 V
51	+8 V		Unregulated supply rail
52	−16 V		Unregulated supply rail
53	GND		Common 0 V
54	SLAVE CLR	L*	Slave clear (resets all bus slaves)
55	DMA0	L*	DMA request line (lowest priority)
56	DMA1	L*	DMA request line
57	DMA2	L*	DMA request line
58	sXTRQ	L*	16-bit data request (requests slaves to assert SIXTN)
59	A19		Address line 19
60	SIXTN	L*	16-bit data acknowledge (slave response to sXTRQ)
61	A20		Extended address bus line 20
62	A21		Extended address bus line 21
63	A22		Extended address bus line 22
64	A23		Extended address bus line 23
65	NDEF		Not defined
66	NDEF		Not defined
67	PHANTOM		Phantom (disables normal slaves and enables phantom slaves which share addresses with the normal set)
68	MWRT	H	Memory write
69	RFU		Reserved for future use

S-100 bus 155

Pin No.	Abbreviation	Active level	Signal/function
70	GND		Common 0 V
71	RFU		Reserved for future use
72	RDY	H*	Ready input to bus master
73	INT	L*	Interrupt request
74	HOLD	L*	Hold request (request from device wishing to have control of the bus)
75	RESET	L*	Reset (resets bus master devices)
76	pSYNC	H	Synchronizing signal which indicates the first bus state of a bus cycle
77	pWR	L	Write (indicates that the bus master has placed valid data on the DO bus/data bus)
78	pDBIN	H	Data bus in (indicates that the bus master is requesting data on the DI bus/data bus)
79	A0		Address line 0
80	A1		Address line 1
81	A2		Address line 2
82	A6		Address line 6
83	A7		Address line 7
84	A8		Address line 8
85	A13		Address line 13
86	A14		Address line 14
87	A11		Address line 11
88	DO2/DATA 2		Data out line 2/bidirectional data line 2
89	DO3/DATA 3		Data out line 3/bidirectional data line 3
90	DO7/DATA 7		Data out line 7/bidirectional data line 7
91	DI4/DATA 12		Data in line 4/bidirectional data line 12
92	DI5/DATA 13		Data in line 5/bidirectional data line 13
93	DI6/DATA 14		Data in line 6/bidirectional data line 14
94	DI1/DATA 9		Data in line 1/bidirectional data line 9
95	DI0/DATA 8		Data in line 0/bidirectional data line 8
96	sINTA	H	Interrupt acknowledge
97	sWO	L	Write output (used to gate data from the bus master to a slave)
98	ERROR	L*	Error (indicates that an error has occurred during the current bus cycle)
99	POC	L	Power on clear (clears all devices attached to the bus when power is first applied)
100	GND		Common 0 V

Notes: 1. * = open collector
  H = active high
  L = active low
 2. p precedes control line mnemonics
   s precedes status line mnemonics

S-100 pin numbering

```
50                                    1
┌─────────────────//─────────────────┐
└─────────────────//─────────────────┘
100                                  51
```

(Pin view of motherboard connector)

Intel Multibus

In recent years Intel's Multibus has gained considerable support as an industrial bus standard. The system is based on an 86-way edge connector and is ideally suited to multiprocessor applications.

The following points are worthy of note:
(a) Signals have been grouped together according to their logical function and placed physically adjacent on the edge connector.
(b) All signals are active low.
(c) 20 address lines and 16 bidirectional data lines are provided (the 20 address lines being those normally associated with the 8086/80186 processor families).
(d) When a processor seeks control of the bus it asserts a bus request signal (BREQ) together with its bus priority output signal (BPRO). A parallel or serial priority encoder uses these signals to determine which processor has control of the bus at any particular time. A suitable bus controller is Intel's 8218 or 8288.
(e) Hexadecimal numbering is used to distinguish address and data bus lines.

Intel Multibus pin assignment

Component side

Pin No.	Signal group	Abbreviation	Signal/function
1	Supply rails	GND	Ground/common 0 V
3		+5 V	+5 V d.c. supply rail
5		+5 V	+5 V d.c. supply rail
7		+12 V	+12 V d.c. supply rail
9		−5 V	−5 V d.c. supply rail
11		GND	Ground/common 0 V
13	Bus control	BCLK	Bus clock
15		BPRN	Bus priority input
17		BUSY	Bus busy
19		MRDC	Memory read command
21		IORC	I/O read command
23		XACK	Transfer acknowledge
25			Reserved
27		BHEN	Byte high enable
29		CBRQ	Common bus request
31		CCLK	Constant clock
33		INTA	Interrupt acknowledge
35	Interrupt	INT6	Parallel interrupt
37		INT4	requests
39		INT2	
41		INT0	
43	Address bus	ADRE	Address lines
45		ADRC	
47		ADRA	
49		ADR8	
51		ADR6	
53		ADR4	

Pin No.	Signal group	Abbreviation	Signal/function
55		ADR2	
57		ADR0	
59	Data	DATE	Data lines
61	bus	DATC	
63		DATA	
65		DAT8	
67		DAT6	
69		DAT4	
71		DAT2	
73		DAT0	
75	Supply	GND	Ground/common 0 V
77	rails		Reserved
79		−12 V	−12 V d.c. supply rail
81		+5 V	+5 V d.c. supply rail
83		+5 V	+5 V d.c. supply rail
85		GND	Ground/common 0 V

Track side

Pin No.	Signal group	Abbreviation	Signal/function
2	Supply	GND	Ground/common 0 V
4	rails	+5 V	+5 V d.c. supply rail
6		+5 V	+5 V d.c. supply rail
8		+12 V	+12 V d.c. supply rail
10		−5 V	−5 V d.c. supply rail
12		GND	Ground/common 0 V
14	Bus	INIT	Initialize
16	control	BPRO	Bus priority output
18		BREQ	Bus request
20		MWTC	Memory write command
22		IOWC	I/O write command
24		INH1	Inhibit 1 (disable RAM)
26		INH2	Inhibit 2 (disable ROM)
28	Address	AD10	Address lines
30	bus	AD11	
32		AD12	
34		AD13	
36	Interrupt	INT7	Parallel interrupt
38		INT5	requests
40		INT3	
42		INT1	
44	Address	ADRF	Address lines
46	bus	ADRD	
48		ADRB	
50		ADR9	
52		ADR7	
54		ADR5	

Pin No.	Signal group	Abbreviation	Signal/function
56		ADR3	
58		ADR1	
60	Data	DATF	Data lines
62	bus	DATD	
64		DATB	
66		DAT9	
68		DAT7	
70		DAT5	
72		DAT3	
74		DAT1	
76	Supply	GND	Ground/common 0 V
78	rails		Reserved
80		−12 V	−12 V d.c. supply rail
82		+5 V	+5 V d.c. supply rail
84		+5 V	+5 V d.c. supply rail
86		GND	Ground/common 0 V

IBM PC expansion bus

The immense popularity of the IBM PC has ensured a wide market for compatible expansion cards and it is therefore not surprising that the IBM PC expansion bus has established itself as a standard in its own right.

The IBM expansion bus uses a double sided 31-way connector (62-ways in total). The address and data bus are grouped together on one side of the connector while the control bus and power rails occupy the other side of the connector.

IBM PC expansion bus pin assignment

Pin No.	Abbreviation	Signal/function
1	GND	Ground/common 0 V
2	CHCK	Channel check output (when low this indicates that some form of error has occurred)
3	RESET	Reset (when high this line resets all expansion cards)
4	D7	Data line 7
5	+5 V	+5 V d.c. supply rail
6	D6	Data line 6
7	IRQ2	Interrupt request input 2
8	D5	Data line 5
9	−5 V	−5 V d.c. supply rail
10	D4	Data line 4
11	DRQ2	DMA request input 2
12	D3	Data line 3

Pin No.	Abbreviation	Signal/function
13	−12 V	−12 V d.c. supply rail
14	D2	Data line 2
15		Reserved
16	D1	Data line 1
17	+12 V	+12 V d.c. supply rail
18	D0	Data line 0
19	GND	Ground/common 0 V
20	BCRDY	Ready input (normally high, pulled low by a slow memory or I/O device to signal that it is not ready for data transfer to take place)
21	$\overline{\text{IMW}}$	Memory write output
22	AEN	Address enable output
23	$\overline{\text{IMR}}$	Memory read output
24	A19	Address line 19
25	$\overline{\text{IIOW}}$	I/O write output
26	A18	Address line 18
27	$\overline{\text{IIOR}}$	I/O read output
28	A17	Address line 17
29	DACK3	DMA acknowledge output 3 (see notes)
30	A16	Address line 16
31	DRQ3	DMA request input 3
32	A15	Address line 15
33	DACK1	DMA acknowledge output 1 (see notes)
34	A14	Address line 14
35	DRQ1	DMA request input 1
36	A13	Address line 13
37	DACK0	DMA acknowledge output 0 (see notes)
38	A12	Address line 12
39	XCLK4	4 MHz clock (CPU clock divided by two, 200 ns period, 50% duty cycle)
40	A11	Address line 11
41	IRQ7	Interrupt request line 7 (see notes)
42	A10	Address line 10
43	IRQ6	Interrupt request line 6 (see notes)
44	A9	Address line 9
45	IRQ5	Interrupt request line 5
46	A8	Address line 8
47	IRQ4	Interrupt request line 4 (see notes)
48	A7	Address line 7
49	IRQ3	Interrupt request line 3
50	A6	Address line 6
51	DACK2	DMA acknowledge 2
52	A5	Address line 5
53	TC	Terminal count output (pulsed high to indicate that the terminal count for a DMA transfer has been reached)
54	A4	Address line 4
55	ALE	Address latch enable output
56	A3	Address line 3
57	+5 V	+5 V d.c. supply rail
58	A2	Address line 2
59	14 MHz	14.31818 MHz clock (fast clock with 70 ns period, 50% duty cycle)

Pin No.	Abbreviation	Signal/function
60	A1	Address line 1
61	GND	Ground/common 0 V
62	IA0	Address line 0

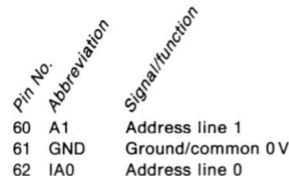

Notes: (a) Signal direction is quoted relative to the motherboard
(b) IRQ4 is generated by the motherboard serial interface
IRQ6 is generated by the motherboard disk interface
IRQ7 is generated by the motherboard parallel interface
(c) DACK0 is used to refresh dynamic memory, while DACK1 to DACK3 are used to acknowledge DMA requests.

IBM PC expansion bus pin numbering

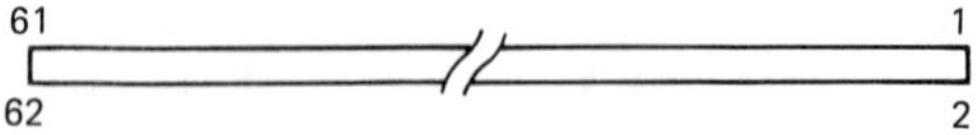

Centronics printer interface

The Centronics interface has established itself as the standard for parallel data transfer between a microcomputer and a printer. The standard is based on 36-way Amphenol connector (part no: 57–30360) and is suitable for distances of up to 2 m.

Parallel data is transferred into the printer's internal buffer when a strobe pulse is sent. Handshaking is accomplished by means of acknowledge (ACKNLG) and busy (BUSY) signals.

Centronics printer interface pin assignment

Pin No.	Abbreviation	Signal/function
1	STROBE	Strobe (active low to read data)
2	DATA 1	Data line 1
3	DATA 2	Data line 2
4	DATA 3	Data line 3
5	DATA 4	Data line 4
6	DATA 5	Data line 5
7	DATA 6	Data line 6
8	DATA 7	Data line 7
9	DATA 8	Data line 8
10	ACKNLG	Acknowledge (pulsed low to indicate that data has been received)
11	BUSY	Busy (taken high under the following conditions: (a) during data entry (b) during a printing operation

Centronics printer interface 161

Pin No.	Abbreviation	Signal/function
		(c) when the printer is OFF-LINE
		(d) during print error status
12	PE	Paper end (taken high to indicate that the printer is out of paper)
13	SLCT	Select (taken high to indicate that the printer is in the selected state)
14	AUTO FEED XT	Automatic feed (when this input is taken low, the printer is instructed to produce an automatic line feed after printing. This function can be selected internally by means of a DIP switch)
15	n.c.	Not connected (unused)
16	0 V	Logic ground
17	CHASSIS GND	Printer chassis (normally isolated from logic ground at the printer)
18	n.c.	Not connected (unused)
19 to 30	GND	Signal ground (originally defined as 'twisted pair earth returns' for pin numbers 1 to 12 inclusive)
31	INIT	Initialize (this line is pulsed low to reset the printer controller)
32	ERROR	Error (taken low by the printer to indicate: (a) PAPER END state (b) OFF-LINE state (c) error state)
33	GND	Signal ground
34	n.c.	Not connected (unused)
35	LOGIC 1	Logic 1 (usually pulled high via 3.3 kohm)
36	SLCT IN	Select input (data entry to the printer is only possible when this line is taken low, but this function may be disabled by means of an internal DIP switch)

Notes: (a) Signals, pin numbers, and signal directions apply to the printer.
(b) Alternative types of connector (such as 25-way D type, PCB edge, etc.) are commonly used at the microcomputer.
(c) All signals are standard TTL levels.
(d) ERROR and ACKNLG signals are not supported on some interfaces.

Centronics interface pin connections

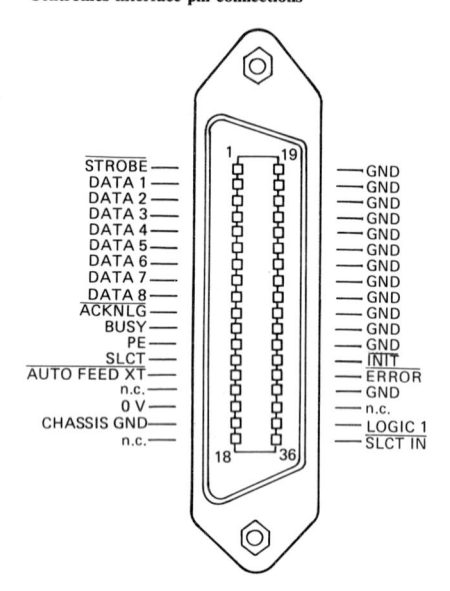

```
STROBE     — 1         19 —  GND
DATA 1     —              —  GND
DATA 2     —              —  GND
DATA 3     —              —  GND
DATA 4     —              —  GND
DATA 5     —              —  GND
DATA 6     —              —  GND
DATA 7     —              —  GND
DATA 8     —              —  GND
ACKNLG     —              —  GND
BUSY       —              —  GND
PE         —              —  GND
SLCT       —              —  INIT
AUTO FEED XT —            —  ERROR
n.c.       —              —  GND
0 V        —              —  n.c.
CHASSIS GND —             —  LOGIC 1
n.c.       — 18        36 —  SLCT IN
```

IEEE-488/GPIB bus

The IEEE-488 bus was originally developed by Hewlett-Packard and is now known simply as the General Purpose Instrument Bus (GPIB). The bus provides a means of interconnecting instruments in automatic test equipment (ATE) configurations where data can be exchanged between a number of participating devices.

The IEEE-488 bus provides for the following types of device:
 (a) Listeners (which receive data from other instruments but are not themselves capable of generating data),
 (b) Talkers (which are only capable of outputing data onto the bus),
 (c) Talker/listeners (which can both generate data and receive data), and
 (d) Controllers (which manage the flow of data on the bus and provide processing facilities).

While only one talker can be active at any given instant, it is possible for several listeners to be receiving data simultaneously.

The IEEE-488 bus uses eight multi-purpose bidirectional data lines. These lines are used to transfer data, addresses, commands and status bytes. In addition, five bus management and three handshake lines are provided.

Commands are signalled by taking the Attention Line (ATN) low. Commands may be directed to individual devices by placing a unique address on the lower five data bus lines. Alternatively, universal commands may be simultaneously directed to all participating devices.

The format for a command byte (ATN low) is as follows:
Bit 8   Don't care
Bits 6 and 7   Command code bits
Bits 5 to 1   Addresses

The following command byte truth table is obeyed:

D8	D7	D6	D5	D4	D3	D2	D1	Meaning
X	0	0	0	B4	B3	B2	B1	Universal commands
X	0	1	A5	A4	A3	A2	A1	Listen addresses
X	0	1	1	1	1	1	1	Unlisten command
X	1	0	A5	A4	A3	A2	A1	Talk addresses
X	1	0	1	1	1	1	1	Untalk command
X	1	1	A5	A4	A3	A2	A1	Secondary commands
X	1	1	1	1	1	1	1	Ignored

Notes: (a) Command codes are only valid when ATN is low.
(b) Address 11111 cannot be used for a talker or listener.

The specified maximum data rate for the bus is normally either 1 or 2 Mbyte/s but, in practice, typical data transfer rates are 250 kbyte/s or less. The maximum terminated bus length is normally assumed to be 20 m.

An important feature of the bus is that it provides for data transfer between devices having widely different response times. In practice, therefore, the slowest listener determines the rate at which data transfer takes place.

A variety of dedicated VLSI GPIB bus interface devices are available. These include Intel's 8291 GPIB Listener/Talker and 8292 GPIB Controller, as well as Motorola's 68488 GPIB Adaptor.

The IEC-625 bus is similar to the IEEE-488 bus but employs a different connector (25-way D-type rather than the 24-pin connector originally specified by Hewlett-Packard).

IEEE-488/GPIB bus system

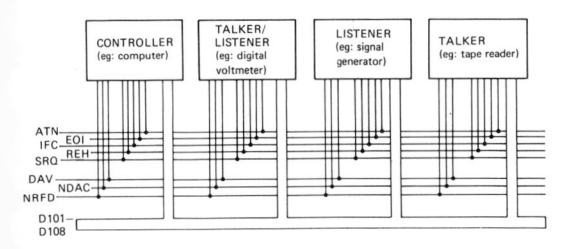

IEEE-488/GPIB pin assignment

Pin No.	Signal group	Abbreviation	Signal/function
1	Data	DIO1	Data line 1
2		DIO2	Data line 2
3		DIO3	Data line 3
4		DIO4	Data line 4

Pin No.	Signal group	Abbreviation	Signal/function
5	Management	EOI	End or identify (sent by a talker to indicate that transfer of data is complete)
6	Handshake	DAV	Data valid (asserted by a talker to indicate that valid data is present on the bus)
7		NRFD	Not ready for data (asserted by a listener to indicate that it is not ready for data)
8		NDAC	Not data accepted (asserted while data is being accepted by a listener)
9	Management	IFC	Interface clear (asserted by the controller in order to initialize the system in a known state)
10		SRQ	Service request (sent to the controller by a device requiring attention)
11		ATN	Attention (asserted by the controller when placing a command onto the bus)
12		SHIELD	Shield
13	Data	DIO5	Data line 5
14		DIO6	Data line 6
15		DIO7	Data line 7
16		DIO8	Data line 8
17	Management	REN	Remote enable (enables an instrument to be controlled by the bus controller rather than by its own front panel controls)
18		GND	Ground/common
19		GND	Ground/common
20		GND	Ground/common
21		GND	Ground/common
22		GND	Ground/common
23		GND	Ground/common
24		GND	Ground/common

Notes: (a) Handshake signals (DAV, NRFD and NDAC) are all active low open collector and are used in a wired-OR configuration.
(b) All other signals are TTL compatible and active low.

IEEE-488/GPIB pin connections

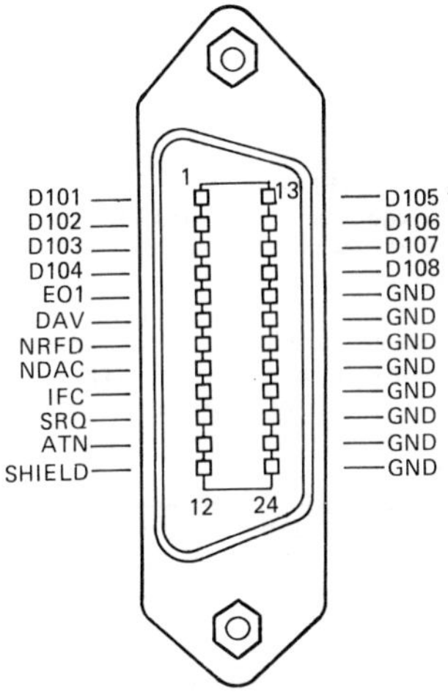

Serial data transmission

In serial data transmission one data bit is transmitted after another. In order to transmit a byte of data it is therefore necessary to convert incoming parallel data from the bus into a serial bit stream which can be transmitted along a line.

Serial data transmission can be synchronous (clocked) or asynchronous (non-clocked). The latter method has obvious advantages and is by far the most popular method. The rate at which data is transmitted is given by the number of bits transmitted per unit time. The commonly adopted unit is the 'baud', with 1 baud roughly equivalent to 1 bit per second.

It should, however, be noted that there is a subtle difference between the bit rate as perceived by the computer and the baud rate present in the transmission medium. The reason is simply that some overhead in terms of additional synchronizing bits is required in order to recover asynchronously transmitted data.

In the case of a typical RS-232C link, a total of 11 bits is required to transmit only seven bits of data. A line baud rate of 600 baud thus represents a useful data transfer rate of only some 382 bits per second.

Many modern serial data transmission systems can trace their origins to the 20 mA current loop interface which was once commonly used to connect a teletype unit to a minicomputer system. This system was based on the following logic levels:

Mark = logic 1 = 20 mA current flowing

Space = logic 0 = no current flowing

where the terms 'mark' and 'space' simply refer to the presence or absence of a current.

This system was extended to cater for more modern and more complex peripherals for which voltage, rather than current, levels were appropriate.

RS-232C/CCITT V24

The RS-232C/CCITT V24 interface is the most widely used method of providing serial communication between microcomputers and peripheral devices. The interface is defined by the Electronic Industries Association (EIA) standard and relates to the connection of data terminal equipment (DTE) and data communication equipment (DCE). For many purposes the DTE and DCE are the computer and peripheral respectively although the distinction is not always clear as, for example, in the case where two microcomputers are linked together via RS-232C ports. In general, the RS-232C system may be used where the DTE and DCE are physically separated by up to 20 m or so. For greater distances telephone lines are usually more appropriate.

The EIA specification permits synchronous or asynchronous communication at data rates of up to 19.2 kb/s. Furthermore, character length and bit codes may be varied according to the particular application. The specification allows for the following signals:

(a) serial data comprising
    (i)   a primary channel providing full duplex data transfer (i.e. simultaneous transmission and reception), and
    (ii)   a secondary channel also capable of full duplex operation;
(b) handshake control signals;
(c) timing signals.

The RS-232C system may thus be configured for a variety of operating modes including transmit only (primary channel), receive only (primary channel), half-duplex, full-duplex, and various primary and secondary channel transmit/receive combinations.

From this, it should be clear that the RS-232C is versatile and highly adaptable. Unfortunately, such flexibility does carry a penalty — the wide variation in interpretation, which can result in some bewildering anomalies in the physical connection and control protocol of practical RS-232C systems.

The RS-232C interface is usually distinguished by its connector — a 25-way 'D' connector. DCE equipment is normally fitted with a female connector while DTE equipment is fitted with a male connector.

In practice, few systems involving personal computers make use of the full complement of signal lines; indeed, many arrangements use only eight lines in total (including the protective ground and signal return).

The most common arrangement for a microcomputer RS-232C interface involves six signal lines and two ground connections. These use pins 1 to 7 and 20 of the D-connector and their functions, assuming that we are dealing with the computer side of the interface, are as follows:

PROTECTIVE GROUND	Connected to the equipment frame or chassis (may be connected to an outer screening conductor).
TD (TxD)	Serial transmitted data output.
RD (RxD)	Serial received data input.
RTS	Request to send. Output. Peripheral to transmit data when an 'on' condition is present.

CTS	Clear to send. Input. When 'on' indicates that the peripheral can receive data.
DSR	Data set ready. Input. When 'on' indicates that handshaking has been completed.
SIGNAL GROUND	Acts as a common signal return. Normally connected to a ground point within the RS-232C interface and should not be linked directly to the protective ground (even though these may appear to both be at zero potential).
DTR	Data terminal ready. Output. When 'on' indicates that the peripheral should be connected to the communication channel.

Serial RS-232C data is transmitted asynchronously (i.e. it is not clocked) and each data word represents a single ASCII character. Most systems provide for seven data bits although some may be configured for any number of bits between five and eight. The number of stop bits may also range between 1, $1\frac{1}{2}$ and 2. Parity may be even, odd or disabled.

The voltage levels in an RS-232C system are markedly different from those which appear within the computer. In the transmit and receive data paths, for example, a positive voltage of between 3 V and 25 V is used to represent logic 0 while a negative voltage of similar magnitude is used to represent logic 1.

In the control signal paths, however, conventional positive logic is employed; a high voltage in the range 3 V to 25 V indicates the active or 'on' state while a negative voltage of similar magnitude indicates the inactive or 'off' state. It should be noted that some 'quasi RS-232C' systems exist in which conventional TTL logic levels are employed. Such systems are obviously not directly compatible with the original EIA system and considerable damage can be caused by inadvertent interconnection of the two.

The maximum open-circuit voltage on any RS-232C line must not be allowed to exceed ±25 V (relative to signal ground) and the maximum short-circuit current between any two lines must not exceed 500 mA. The effective loading resistance of any circuit must be between 3 kohm and 7 kohm with an effective shunt capacitance not exceeding 2.5 nF.

RS-423/RS-422

In order to improve the performance of the RS-232C specification, several further standards have been introduced. These provide for better line matching, thereby reducing reflections which are otherwise present on a mismatched line.

RS-422 is a balanced system (differential signal lines are used) while RS-423 is unbalanced (a single signal line is used in conjunction with signal ground). Both systems allow several remote peripherals to be driven from a common line in a simple serial bus configuration.

RS-423 allows for a line terminating resistance of 450 ohm (minimum) while RS-422 caters for a line impedance of as low as 50 ohm. The improvement in matching permits the use of much faster data rates. Data rates of up to 125 kbaud and distances of over 1000 m can be tolerated (though not necessarily at the same time!).

RS-449

RS-449 is a further enhancement of RS-422 and RS-423 which has been developed to cater for very fast data rates (up to 2 Mbaud). Compared with RS-232C, ten extra circuit functions have been provided while three of the original interchange circuits have been abandoned. Protective ground is also no longer provided.

In order to minimize confusion, and since certain changes have been made to the definition of circuit functions, a completely new set of mnemonics has been provided. In addition, the system requires 37-way and 9-way 'D' connectors, the latter being necessary where use is made of the secondary channel interchange circuits.

RS-232C/CCITT V24 pin assignment

Pin No.	Abbrev.	Direction	Circuit CCITT	Circuit EIA	Function
1	FG	—	101	AA	Frame ground
2	TD	To DCE	103	BA	Transmitted data
3	RD	To DTE	104	BB	Received data
4	RTS	To DCE	105	CA	Request to send
5	CTS	To DTE	106	CB	Clear to send
6	DSR	To DTE	107	CC	Data set ready
7	SG	—	102	AB	Signal ground
8	DCD	To DTE	109	CF	Data carrier detect
9		To DTE			Positive d.c. test voltage
10		To DTE			Negative d.c. test voltage
11	QM	To DTE	Note 1		Equalizer mode
12	SDCD	To DTE	122	SCF	Secondary data carrier detect
13	SCTS	To DTE	121	SCB	Secondary clear to send
14	STD	To DCE	118	SBA	Secondary transmitted data
	NS	To DCE	Note 1		New synchronization
15	TC	To DTE	114	DB	Transmitter clock
16	SRD	To DTE	119	SBB	Secondary received data
	DCT	To DTE	Note 1		Divided clock transmitter
17	RC	To DTE	115	DD	Receiver clock
18	DCR	To DTE	Note 1		Divided clock receiver
19	SRTS	To DCE	120	DCA	Secondary request to send
20	DTR	To DCE	108.2	CD	Data terminal ready
21	SQ	To DTE	110	CG	Signal quality detect
22	RI	To DTE	125	CE	Ring indicator
23		To DCE	111	CH	Data rate selector
		To DCE	112	CI	Data rate selector
24	TC	To DCE	113	DA	External transmitter clock
25		To DCE	Note 2		Busy

Notes: 1. Bell 208A
2. Bell 113B

RS-232C pin connections

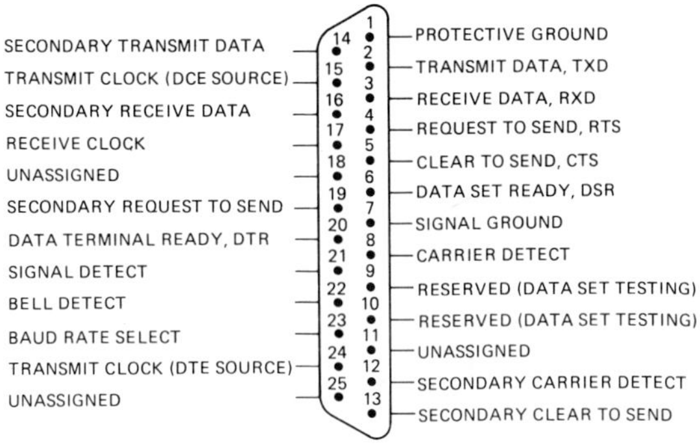

Pin	Signal
1	PROTECTIVE GROUND
2	TRANSMIT DATA, TXD
3	RECEIVE DATA, RXD
4	REQUEST TO SEND, RTS
5	CLEAR TO SEND, CTS
6	DATA SET READY, DSR
7	SIGNAL GROUND
8	CARRIER DETECT
9	RESERVED (DATA SET TESTING)
10	RESERVED (DATA SET TESTING)
11	UNASSIGNED
12	SECONDARY CARRIER DETECT
13	SECONDARY CLEAR TO SEND
14	SECONDARY TRANSMIT DATA
15	TRANSMIT CLOCK (DCE SOURCE)
16	SECONDARY RECEIVE DATA
17	RECEIVE CLOCK
18	UNASSIGNED
19	SECONDARY REQUEST TO SEND
20	DATA TERMINAL READY, DTR
21	SIGNAL DETECT
22	BELL DETECT
23	BAUD RATE SELECT
24	TRANSMIT CLOCK (DTE SOURCE)
25	UNASSIGNED

Serial data format

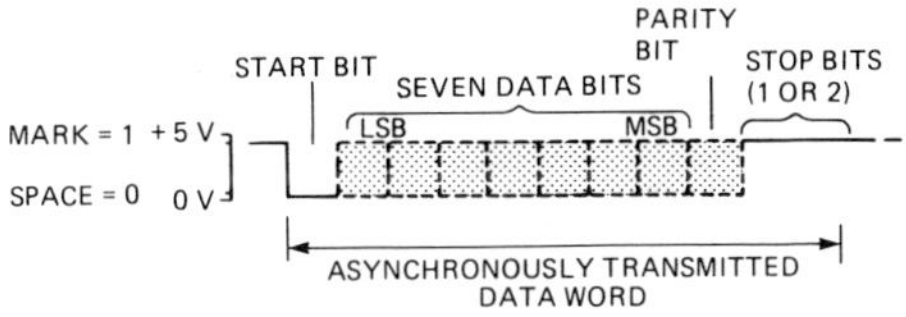

(a) TTL levels

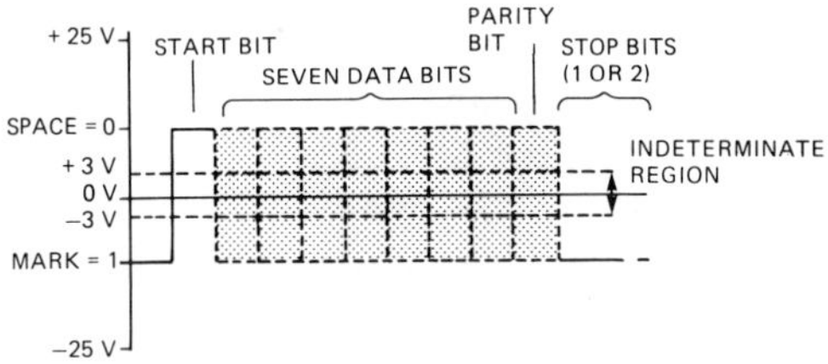

MARK = 1 +5 V
SPACE = 0   0 V

START BIT, SEVEN DATA BITS (LSB ... MSB), PARITY BIT, STOP BITS (1 OR 2)

ASYNCHRONOUSLY TRANSMITTED DATA WORD

(b) RS-232C signal levels

+25 V
SPACE = 0
+3 V
0 V
−3 V
MARK = 1
−25 V

START BIT, SEVEN DATA BITS, PARITY BIT, STOP BITS (1 OR 2), INDETERMINATE REGION

Data rates and distances for RS-232C

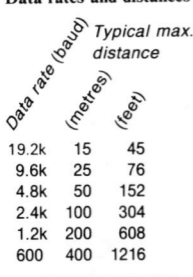

Data rate (baud)	Typical max. distance (metres)	(feet)
19.2k	15	45
9.6k	25	76
4.8k	50	152
2.4k	100	304
1.2k	200	608
600	400	1216

RS-232C/CCITT V24 voltage levels

Mark = logic 1 = 'off' = -3 V to -25 V (typically -6 V)
Space = logic 0 = 'on' = $+3$ V to $+25$ V (typically $+6$ V)

170 RS-232C/CCITT V24

Null modems

SIGNAL	PIN No	PIN No
FG	1 — 1
TD	2 ⨯ 2
RD	3 ⨯ 3
RTS	4 ⨯ 4
CTS	5 ⨯ 5
DSR	6 — 6
SG	7 — 7
DCD	8 — 8
DTR	20 — 20

SIGNAL	PIN No	PIN No
FG	1 — 1
TD	2 ⨯ 2
RD	3 ⨯ 3
RTS	4 — 4
CTS	5 — 5
DSR	6 ⨯ 6
SG	7 — 7
DCD	8 ⨯ 8
DTR	20 — 20

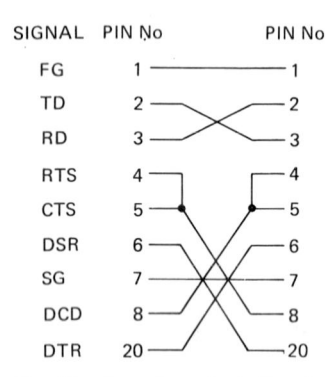

Note: The above alternative 'null modems' may be used to link two computers together where each device is configured as a DTE.

Typical serial data communication support devices

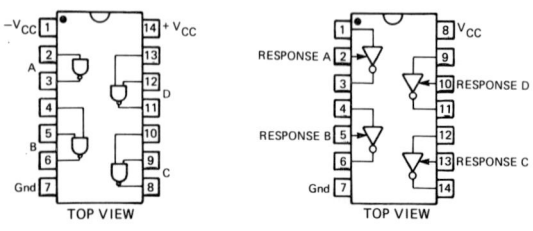

1488 RS-232C Line Driver

1489 RS-232C Line Receiver

RS-232C/CCITT V24 171

Typical serial data communication support devices *cont.*

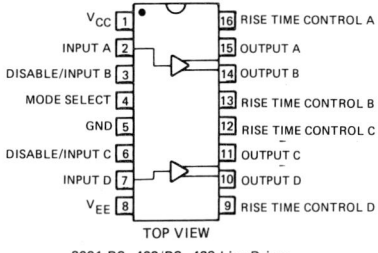

```
        V_CC  [1  •        16]  RISE TIME CONTROL A
     INPUT A  [2           15]  OUTPUT A
DISABLE/INPUT B [3         14]  OUTPUT B
 MODE SELECT  [4           13]  RISE TIME CONTROL B
         GND  [5           12]  RISE TIME CONTROL C
DISABLE/INPUT C [6         11]  OUTPUT C
     INPUT D  [7           10]  OUTPUT D
        V_EE  [8            9]  RISE TIME CONTROL D
```
TOP VIEW
3691 RS-422/RS-423 Line Driver

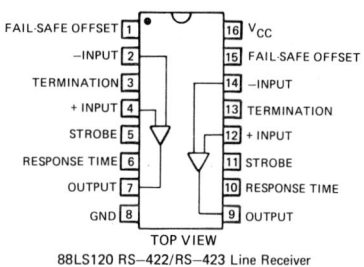

```
FAIL-SAFE OFFSET [1  •     16]  V_CC
        -INPUT  [2         15]  FAIL-SAFE OFFSET
   TERMINATION  [3         14]  -INPUT
        +INPUT  [4         13]  TERMINATION
        STROBE  [5         12]  +INPUT
 RESPONSE TIME  [6         11]  STROBE
        OUTPUT  [7         10]  RESPONSE TIME
           GND  [8          9]  OUTPUT
```
TOP VIEW
88LS120 RS-422/RS-423 Line Receiver

Typical baud rate generator circuit

74LS259 output	Baud rate
Q_0	110
Q_1	9600
Q_2	4800
Q_3	1800
Q_4	1200
Q_5	2400
Q_6	300
Q_7	150

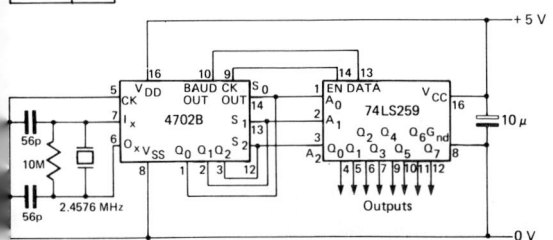

Modem standards

Tone frequencies (Hz)

Classification	Data rate (baud)	Mode	Transmit 0	Transmit 1	Receive 0	Receive 1	Answer
CCITT V21 Orig.	300	Duplex	1180	980	1850	1650	—
CCITT V21 Answ.	300	Duplex	1850	1650	1180	980	2100
CCITT V23 Mode 1	600	Half duplex	1700	1300	1700	1300	2100
CCITT V23 Mode 2	1200	See note	2100	1300	2100	1300	2100
CCITT V23 Back	75	See note	450	390	450	390	—
Bell 103 Orig.	300	Duplex	1070	1270	2025	2225	—
Bell 103 Answ.	300	Duplex	2025	2225	1070	1270	2225
Bell 202	1200	Half	2200	1200	2200	1200	2025

Note: V23 Mode 2 permits full-duplex when the 'back' data rate is 75 baud. When the 'back' rate is 1200 baud the system operates in half-duplex.

S5/8 interface

The RS-232C and V.24 standards are unnecessarily complex for many applications and a simpler serial interface using conventional TTL levels has much to recommend it. A minimal but nevertheless elegant solution is offered by the new S5/8 standard. This standard (currently awaiting BS approval) uses 5 V levels in conjunction with a standard 8-pin DIN connector.

The S5/8 standard specifies two classes of device. A D-device incorporates its own power supply and can provide power (+5 V at up to 20 mA) at the S5/8 connector. An S-device, on the other hand, does not have a supply of its own but may derive its power from an associated D-device. A typical example of an S-device connected to a D-device would be a line-powered modem connected to a personal computer.

Although there is an obvious difference between D and S-devices as regards power supplies, they are considered to be on an equal footing as far as data transfer is concerned; neither device is considered to be a sender or receiver (a perennial bugbear of RS-232C systems!).

The pin assignment of the standard 8-way DIN connector used by S5/8 is as follows:

Pin No.	Abbreviation	Signal/function
1	DINP	Data input
2	GROUND	Signal ground (common)
3	DOUT	Data output
4	HINP	Handshake input
5	HOUT	Handshake output
6	SINP	Secondary input
7	SOUT	Secondary output
8	V+	+5 V (20 mA max.)
Screen	EARTH	Earthed screen

The above arrangement ensures that input and output signals are paired on opposite sides of the connector (as in audio practice). It should also be noted that a standard 180 degree 5-pin DIN plug will mate with the 8-pin DIN connector specified in S5/8. This arrangement will give access to all signals with the exception of the secondary communication circuits (SINP and SOUT) and +5 V (V+).

The electrical characteristics of the S5/8 interface are as follows:

Inputs

Input resistance:	47 k ohm
Input low threshold:	+0.9 V maximum
Input high threshold:	+3.85 V minimum
Input protection:	±25 V minimum

Outputs

Output low voltage:	+0.15 V maximum
Output high voltage:	+4.35 V minimum
Capacitive load drive capability:	2.5 nF minimum
Short-circuit protection:	to any other signal on the interface

S5/8 uses a conventional serial data structure and, in the same sense as RS-232C, the line rests low (0 V) and goes high for the start bit. Thereafter, transmitted data bits are inverted. Each frame (serially transmitted data word) comprises one start bit, eight data bits, and one stop bit (i.e., 10 bits total). There is no parity bit and hence error detection should be performed on a block-by-block basis using checksum or CRC techniques.

S5/8 specifies a data transfer rate of 9600 bit/s (the fastest widely used bit rate) and simple handshaking is provided using the HINP and HOUT lines. A full software flow-control specification is currently awaited.

Undoubtedly the most attractive feature of S5/8 is that an interface can be very easily realized using nothing more than a UART and a high-speed CMOS inverting buffer (e.g. 74HC14). There is no need for the line drivers and level shifters that would be essential to the correct operation of a conventional RS-232C interface.

Finally, whilst an S5/8 input will safely and correctly receive RS-232C signal levels, the reverse is not necessarily true. Depending upon the popularity of the new standard, it is expected that many manufacturers will not only implement their RS-232C ports using 8-way DIN connectors but will also change their line receivers for high-speed CMOS Schmitt inverters. Such an arrangement should readily permit interworking of the two systems.

KERMIT

The KERMIT file transfer protocol was developed by Bill Catchings and Frank da Cruz at the Columbia University Centre for Computing Activities (CUCCA). The initial objective was to allow users of DEC-20 and IBM timesharing systems to archive their files on microcomputer floppy disks. The design owes much to the ANSI and ISO models and ideas were incorporated from similar projects at Stanford University and the University of Utah.

KERMIT has grown to support over fifty different operating systems and is now in constant use in many sites all over the world. The KERMIT software is free and available to all but, to defray costs of media, printing, postage, etc., a distribution fee is requested from sites that order KERMIT directly from the University of Columbia. Other sites are, however, free to distribute KERMIT on their own terms subject to certain stipulations.

Further details can be obtained from:

KERMIT distribution,
Columbia University Centre for Computing Activities,
7th Floor Watson Laboratory,
612 West 115th Street,
New York,
NY 10025

Prospective microcomputer KERMIT users should note that CUCCA can only provide 9-inch tapes (usually 1600 bit/in). Bootstrapping procedures are, however, provided to allow microcomputer versions to be downloaded from the mainframe for which the tape is produced. The tape includes all source programs and documentation. One copy of the KERMIT manual is also provided with each tape.

KERMIT is designed for the transfer of sequential files over ordinary serial telecommunication lines. It is not necessarily better than many other terminal-oriented file transfer protocols but it is free, well documented, and has been implemented on a wide variety of microcomputers and mainframes.

KERMIT transfers data by encapsulating it in 'packets' of control information which incorporate a synchronization marker, packet number (to facilitate detection of 'lost' packets), length indicator, and a checksum to allow verification of the data. Retransmission is requested when lost or corrupt data packets are detected; duplicate packets are simply discarded. In addition, special control packets allow co-operating KERMITs to connect and disconnect from each other and to exchange various kinds of information. Very few assumptions are made concerning the capabilities of either of the participating computers and hence the KERMIT protocol is effective with many different types of system.

KERMIT uses a simple set of basic commands which include SEND (followed by a filespec), RECEIVE, CONNECT (i.e., establish a virtual terminal connection to the remote system), SET (establish non-standard settings such as parity and flow-control), and HELP (displays a summary of KERMIT commands and actions). A ? typed anywhere within a KERMIT command lists the commands, options, or operands that are possible at that point. This particular command may, or may not, require a carriage return depending upon the operating system employed.

Useful interface circuits

Bipolar transistor relay driver

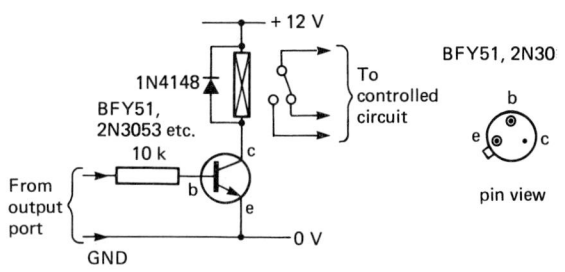

A logic 1 from the output port operates the relay.
Max. recommended relay operating current = 50 mA

VMOS FET relay driver

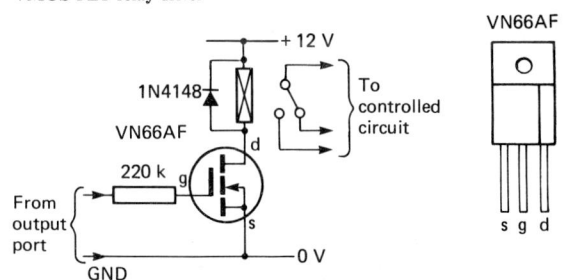

A logic 1 from the output port operates the relay.
Max. recommended relay operating current = 500 mA

VMOS FET motor driver

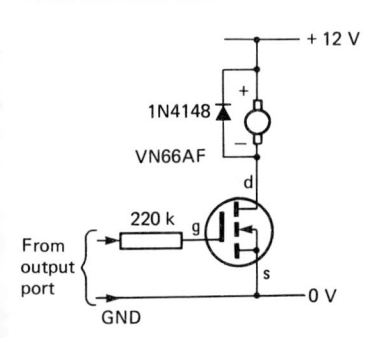

A logic 1 from the output port operates the motor.
Max. recommended motor current (stalled) = 1 A
Max. recommended motor current (operating) = 500 mA

176 Useful interface circuits

VMOS FET audible transducer driver

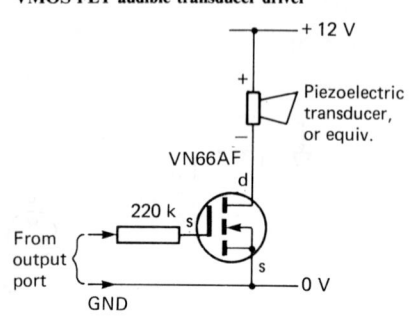

A logic 1 from the output port produces an audible output.
Max. recommended transducer current (operating) = 500 mA

A.C. mains controller using a solid state relay

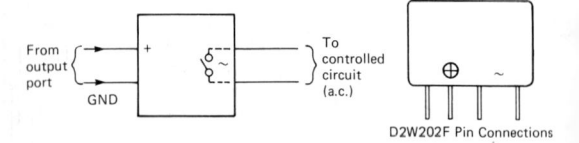

D2W202F Pin Connections

A logic 1 relay will switch the main circuit 'on'.
Typical solid state controller input resistance = 1.5 kohm
(It will therefore interface directly with most TTL devices)
Controlled voltages can be between 60 V and 280 V a.c. at up to 2 A.
Max. 'off' state leakage current = 5 mA
Max. isolation = 2.5 kV a.c.
(The D2W202F is available from International Rectifier)

Bipolar transistor LED driver

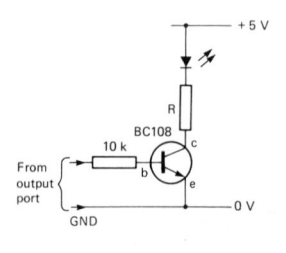

R	Typical diode current
220 Ω	13 mA
270 Ω	10 mA
330 Ω	8.5 mA
390 Ω	7 mA

BC108

pin view

Switch input

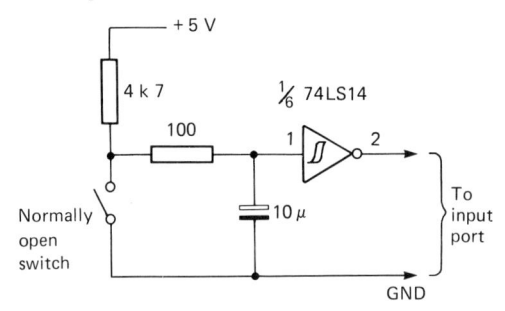

Switch 'open' generates a logic 0 at the input port.
Switch 'closed' generates a logic 1 at the input port.
The circuit is unsuitable for very noisy switches (i.e. where contact bounce is severe) in which case additional software 'de-bouncing' will be required.

Optically isolated data coupler

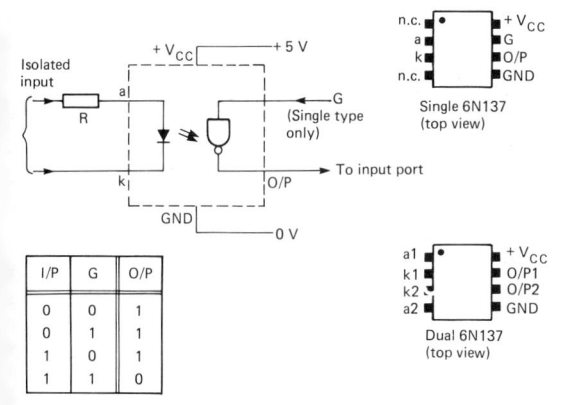

I/P	G	O/P
0	0	1
0	1	1
1	0	1
1	1	0

This interface allows data to be transferred between two electrically isolated systems.
A suitable isolator is the 6N137 (also available in a dual version).
Typical propagation delay = 45 ns
Max. data transfer rate = 10 Mbit/s
Input current = 10 mA typical (therefore necessitating a TTL buffer driver)
Typical value for R = 120 ohm (TTL logic level input)
The enable input (G) is normally held high.

178 Useful interface circuits

Photodiode light sensor interface

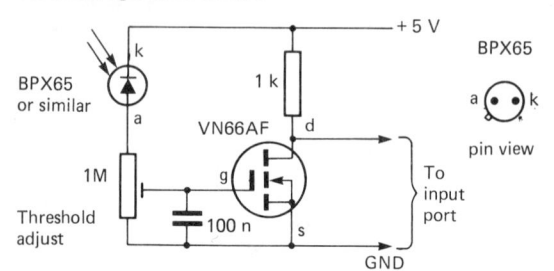

A logic 0 is generated when the light level exceeds the threshold setting, and vice versa.

Semiconductor temperature sensor interface

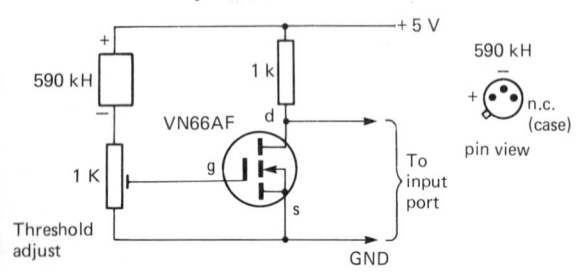

A logic 0 is generated when the temperature level exceeds the threshold setting, and vice versa.

Stepper motor interface

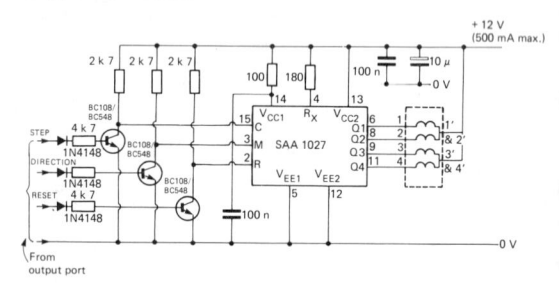

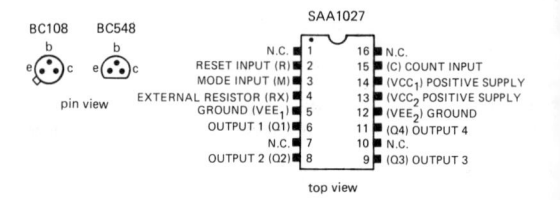

Pin	SAA1027
1	N.C.
2	RESET INPUT (R)
3	MODE INPUT (M)
4	EXTERNAL RESISTOR (RX)
5	GROUND (VEE1)
6	OUTPUT 1 (Q1)
7	N.C.
8	OUTPUT 2 (Q2)
9	(Q3) OUTPUT 3
10	N.C.
11	(Q4) OUTPUT 4
12	(VEE2) GROUND
13	(VCC2) POSITIVE SUPPLY
14	(VCC1) POSITIVE SUPPLY
15	(C) COUNT INPUT
16	N.C.

top view

Stepper motor connections

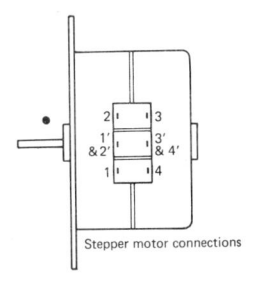

The interface is suitable for driving a four-phase two-stator stepper motor having the following characteristics:

Supply voltage = 12 V
Resistance per phase = 47 ohm
Inductance per phase = 400 mH
Max. working torque = 50 mNm
Step rotation = 7.5 degrees per step

The STEP input is pulsed low to produce a step rotation.

A low (logic 0) on the DIRECTION input selects clockwise rotation. A high (logic 1) on the DIRECTION input selects anticlockwise rotation.

The RESET input is taken high to reset the driver. During normal operation the RESET input must be held low (logic 0).

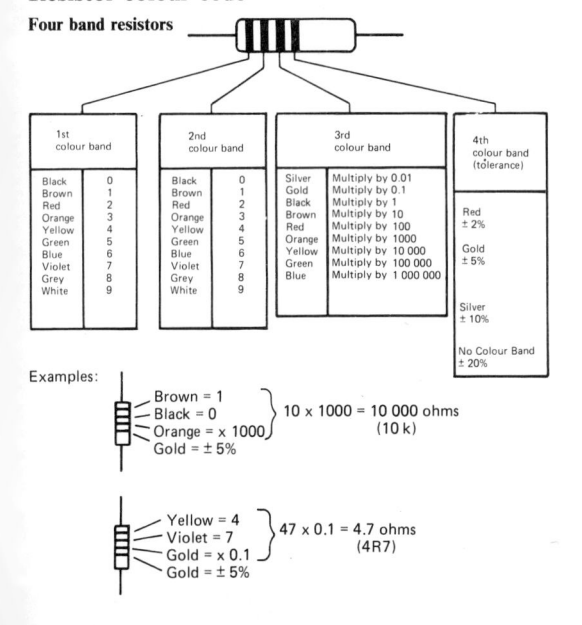

Resistor colour code

Four band resistors

1st colour band		2nd colour band		3rd colour band		4th colour band (tolerance)
Black	0	Black	0	Silver	Multiply by 0.01	
Brown	1	Brown	1	Gold	Multiply by 0.1	
Red	2	Red	2	Black	Multiply by 1	Red ± 2%
Orange	3	Orange	3	Brown	Multiply by 10	
Yellow	4	Yellow	4	Red	Multiply by 100	Gold ± 5%
Green	5	Green	5	Orange	Multiply by 1000	
Blue	6	Blue	6	Yellow	Multiply by 10 000	
Violet	7	Violet	7	Green	Multiply by 100 000	Silver ± 10%
Grey	8	Grey	8	Blue	Multiply by 1 000 000	
White	9	White	9			No Colour Band ± 20%

Examples:

Brown = 1
Black = 0
Orange = × 1000
Gold = ± 5%

10 × 1000 = 10 000 ohms (10 k)

Yellow = 4
Violet = 7
Gold = × 0.1
Gold = ± 5%

47 × 0.1 = 4.7 ohms (4R7)

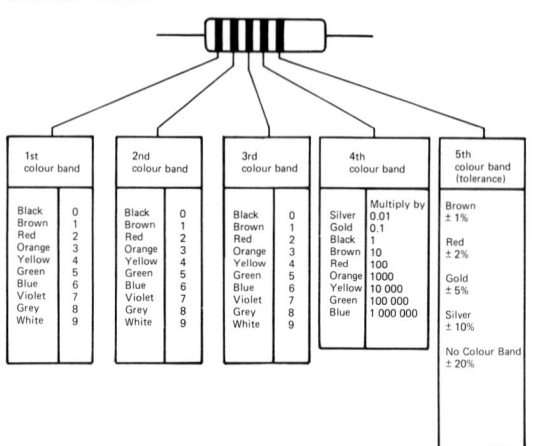

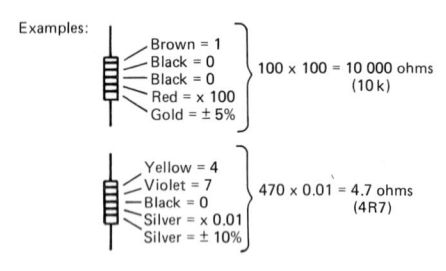

180 Resistor colour code

Five band resistors

1st colour band	2nd colour band	3rd colour band	4th colour band	5th colour band (tolerance)
Black 0	Black 0	Black 0	Multiply by	Brown ± 1%
Brown 1	Brown 1	Brown 1	Silver 0.01	
Red 2	Red 2	Red 2	Gold 0.1	Red ± 2%
Orange 3	Orange 3	Orange 3	Black 1	
Yellow 4	Yellow 4	Yellow 4	Brown 10	Gold ± 5%
Green 5	Green 5	Green 5	Red 100	
Blue 6	Blue 6	Blue 6	Orange 1000	Silver ± 10%
Violet 7	Violet 7	Violet 7	Yellow 10 000	
Grey 8	Grey 8	Grey 8	Green 100 000	No Colour Band ± 20%
White 9	White 9	White 9	Blue 1 000 000	

Examples:

Brown = 1
Black = 0
Black = 0
Red = x 100
Gold = ± 5%

100 × 100 = 10 000 ohms (10 k)

Yellow = 4
Violet = 7
Black = 0
Silver = x 0.01
Silver = ± 10%

470 × 0.01 = 4.7 ohms (4R7)

Capacitor colour code

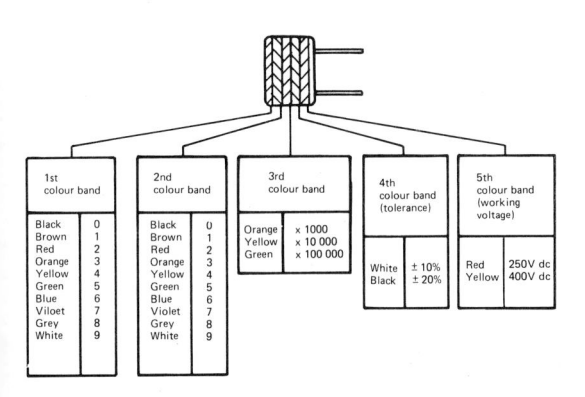

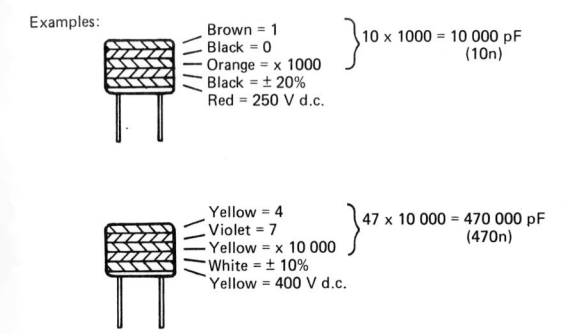

1st colour band	2nd colour band	3rd colour band	4th colour band (tolerance)	5th colour band (working voltage)
Black 0 Brown 1 Red 2 Orange 3 Yellow 4 Green 5 Blue 6 Viloet 7 Grey 8 White 9	Black 0 Brown 1 Red 2 Orange 3 Yellow 4 Green 5 Blue 6 Violet 7 Grey 8 White 9	Orange × 1000 Yellow × 10 000 Green × 100 000	White ± 10% Black ± 20%	Red 250V dc Yellow 400V dc

Examples:

Brown = 1
Black = 0
Orange = × 1000
Black = ± 20%
Red = 250 V d.c.

10 × 1000 = 10 000 pF (10n)

Yellow = 4
Violet = 7
Yellow = × 10 000
White = ± 10%
Yellow = 400 V d.c.

47 × 10 000 = 470 000 pF (470n)

Cassette drive stock faults

Symptom	Cause
Motor not turning	Power supply defective
	Motor control relay/series transistor
	Defective speed regulator
	Defective cable or connector
	Motor defective
Motor turning but tape not moving	Drive belt broken or excessively worn
No read or write	Power supply failure
	Read/write head open circuit
	I/O failure
	Defective cable or connector
	Read/write head dirty or worn
Read but no write (previous data not erased)	Erase head open-circuit
	Erase circuitry not functioning
	Erase head dirty or worn
Read but no write (previous data erased)	Write amplifier defective
Write but no read (tape can be read on a similar machine)	Read amplifier defective
Inability to read tapes made on other machines	Speed incorrect
	Worn drive belt
	Incorrect reading level
Intermittent data errors	Speed irregular
	Worn drive belt
	Worn or eccentric pinch wheel
	Insufficient pinch wheel pressure
	Flywheel defective
	Insufficient erase current
	Incorrect recording or reading levels
	Faulty ALC circuitry
Intermittent data errors with particular media	Inconsistent magnetic oxide coating
	Tape stretched or creased
	Cassette pressure pad defective
Gradual deterioration in performance, increasing number of data errors	Heads contaminated with oxide
	Residual magnetism in read/write head
	Incorrect azimuth of read/write head

Action

Check supply rail
Replace relay/series transistor

Check regulator
Check for d.c. at appropriate points, check continuity with multimeter
Replace motor
Replace belt

Check supply rails and regulators
Check head for continuity, check head connections to PCB
Check I/O device, including address decoding and chip select signals
Check using multimeter or logic probe at appropriate points
Check read/write head, clean or replace

Check head for continuity, check head connections to PCB
Check erase oscillator or d.c. feed to erase head
Check erase head, clean or replace
Check for write signal at the read/write head and work backwards to the I/O device
Check for read signal at the read/write head and work forwards to the I/O device
Check motor speed regulator, adjust if possible

Replace belt
Adjust gain control
Check motor speed regulator
Replace belt
Replace pinch wheel

Adjust pinch wheel

Check for excessive play on flywheel
Check erase signal or d.c. feed to erase head
Check levels and, if possible, adjust

Check ALC detector and time constant
Replace cassette

Replace cassette
Adjust or replace pressure pad

Clean heads

De-magnetize read/write head

Adjust head azimuth

Symptom	Cause
Wind or rewind too slow	Cassette defective
	Friction plate worn
	Friction lever worn
	Torque low
Tapes damaged	Excessive take-up torque
	Excessive pinch wheel pressure
Tape spillage	Pinch wheel not disengaging correctly
	Friction plates worn
Pause inoperative	Pinch wheel not disengaging
Eject inoperative	Eject linkage or eject spring faulty

Disk drive stock faults

Symptom	Cause
Drive motor not turning	Power supply defective
	'Motor on' signal not active
	Defective speed regulator
	Defective cable or connector
	Motor defective
Motor turning but disk not moving	Drive belt broken or excessively worn
	Head load bail arm defective
No read or write	Power supply failure
	Head not loading
	Head not stepping
	Pressure pad assembly defective
	Index hole not located
	Read amplifier defective
	Read/write head open circuit
	Drive not selected
	Read/write head dirty or worn
Read but no write	Disk write-protected
	Write-protect circuitry faulty
	Write amplifier defective

Disk drive stock faults 185

Action

Replace cassette
Replace or adjust friction plate
Replace or adjust lever
Check and adjust pulley assembly
Adjust or replace assembly
Adjust pinch wheel pressure

Check pinch wheel assembly

Replace or renew friction plates
Check pinch wheel assembly and slider linkage

Check linkage or replace spring

Action

Check supply rail
Check disk controller and disk bus (pin 16)

Check regulator
Check for d.c. at appropriate points

Replace motor
Replace belt

Check arm and adjust or replace

Check supply rails and regulators
Check head load mechanism and solenoid.
Check disk bus (pin 4) and work towards solenoid driver
Check stepper motor mechanism. Check disk bus (pin 20) and work towards stepper motor drivers
Check pressure pad and spring tension. Renew if necessary
Check LED and photo-detector circuitry
Check read amplifier and move towards disk controller. Check disk bus (pin 30)
Check head for continuity, check head connections to PCB
Check drive select lines using logic probe
Check read/write head, clean or replace

Remove write-protect tab
Check write-protect LED and photo-detector. Check disk bus (pin 26)
Check for write signal at the read/write head and work backwards to the disk controller.

Symptom	Cause
Inability to read disks made on other machines Intermittent data errors	Disk controller failure Speed incorrect Disk format incorrect Speed irregular Worn drive belt Pressure pad worn Data separation fault
Intermittent data errors with particular disks	Flywheel defective Incorrect recording or reading levels Read/write head dirty or worn Inconsistent magnetic oxide coating Disk hub rings damaged or off-centre Excessive internal friction between disk and envelope
Gradual deterioration in performance, increasing number of data errors	Read/write head contaminated with oxide Read/write head worn Head assembly out of alignment Head carriage worn Insufficient head pressure Speed incorrect
Disk damaged	Excessive head pressure Head worn or damaged Drive spindle and platen out of alignment Foreign body lodged in pressure pad

Printer stock faults

Symptom	Cause
Printer non-functional. Controls and indicators inoperative	Mains input fuse blown Mains switch defective Input filter defective Mains transformer open-circuit

Printer stock faults 187

Action

Check disk bus (pin 20)
Check disk bus (pin 24)
Check motor speed and adjust
Drive and/or DOS incompatible
Check motor speed regulator
Replace belt
Adjust or renew pressure pad
Check data separation circuitry using oscilloscope and test disk
Check for excessive play on flywheel
Check read/write amplifiers

Check read/write head, clean or replace

Replace disk

Replace disk

Replace disk

Clean head

Replace head
Check head azimuth with analogue alignment disk
Adjust or replace head carriage
Adjust pressure pad
Check motor speed and adjust. Soak test and measure speed after drive has reached its normal working temperature
Adjust pressure pad
Replace head
Re-align

Replace pressure pad

Action

Check and replace. If fuse still blown check input filter, mains transformer, and power supply
Disconnect from mains supply and test mains switch for continuity
Check filter inductors for continuity
Check resistance of winding with an ohmmeter. (Typical values of primary and secondary winding resistance are 40 ohm and 1 ohm respectively)

Printer stock faults

Symptom	Cause
	Power supply defective
Head carriage moves but no characters are printed	Head driving pulse absent or too narrow
Head carriage moves but printing is faint or inconsistent	Incorrect head gap
	Incorrect head gap
	Worn ribbon
Head carriage moves but one or more of the dot positions is missing	Defective head driver transistor or open circuit print head
	Defective print head
	Defective driver or buffer
Head carriage does not move, 'out of paper' indicator is illuminated	Paper end detector faulty
Head carriage does not move or moves erratically	Timing belt broken or worn
	Timing belt tension incorrect
	Timing sensor defective
	Defective carriage motor or driver transistor

Action

Check individual raw d.c. rails. Check rectifiers and regulators

Check the head driving pulse using an oscilloscope
Check the head pulse monostable and/or the head trigger from the master CPU
Check positive supply rail to head driver transistors
Check and adjust
Check and adjust
Replace ribbon
Check the waveform at the collector and base of each driver transistor. If any collector waveform is found to be permanently high while the base is normal, remove the transistor, test and replace. If any collector waveform is found to be permanently low while the base is normal, disconnect the print head ribbon cable and measure the resistance of the actuator solenoid in question (typically 22 ohm)
Replace the print head if the actuator solenoid is found to be open circuit, otherwise remove the driver transistor, test and replace
If all waveforms are normal and all actuator solenoids measure approximately 22 ohm, it is possible that one or more of the needles has become seized or broken. It will then be necessary to remove and replace the print head. A substitution test should thus be carried out
If one or more of the base waveforms is incorrect, check the driver using an oscilloscope and work backwards to the master CPU
Check PE signal and paper end sensor

Check and replace

Adjust tension plate assembly

Check PTS signal. Check position timing sensor
Check waveforms at the collector of the four carriage motor driver transistors and verify the correct phase relationship. If one of the collector waveforms is permanently high whilst the base waveform is normal, remove, test and replace the transistor in question. If one of the collector waveforms is permanently low while the base waveform is normal, check the resistance of the relevant winding on the stepper motor (typically 40 to 50 ohm). If necessary compare with values obtained from the other windings. Remove and replace the carriage motor if any one of the windings is

Symptom	Cause
	Defective driver or buffer
Paper feed abnormal or not feeding at all	Defective paper release mechanism or sprocket drive
	Defective line feed motor or driver transistor
	Defective driver or buffer
Printer executes 'self-test' but will not accept printing instructions from the host computer	Interface faulty
Abnormal indication on switch panel. 'LF', 'FF' or 'OFF-LINE' switches inoperative	Control switch or indicator defective

Monitor stock faults

Symptom	Cause
No raster displayed, controls inoperative	Power supply failure
	Horizontal output stage failure

Action

found to be abnormal, otherwise check the driver transistor for a collector-emitter short-circuit

If one or more of the base waveforms is incorrect, check the driver using an oscilloscope and work backwards to the slave CPU

Check friction feed and sprocket drive assembly. Adjust or repla

Check waveforms at the collector and base of each of the four line feed motor driver transistors and verify the correct phase relationship. If one of the collector waveforms is permanently high while the base waveform is normal, remove, test and replace the transistor in question. If one of the collector waveforms is permanently low while the base is normal, check the resistance of the relevant winding on the line feed stepper motor (typically 40 to 50 ohm). If necessary, compare with values obtained from the other windings. Remove and replace the line feed motor if any of the windings is found to be abnormal, otherwise check the driver transistor for a collector-emitter short-circuit

If one or more of the base waveforms is incorrect, check the driver with an oscilloscope and work backwards to the CPU

Check interface cable and connectors. Check interface circuitry and, in particular, the BUSY (pin 11) and $\overline{ERROR}$ (pin 32) status signal lines

Remove and check switch panel. Check switch panel connector and interconnecting cable to main PCB

Clean or replace any defective switch

Test and replace any defective indicator

Action

Check fuses, d.c. supply rails, mains transformer windings for continuity, rectifiers and regulators

Check supply rail to horizontal output stage. Check waveform at the collector of the driver stage and at the collector of the horizontal output stage. If the former is normal while the latter is abnormal, remove and test the output transistor. Replace if defective, otherwise check

Monitor stock faults

Symptom	Cause
	Horizontal oscillator or driver faulty
	CRT defective
Raster is displayed but no video information is present	Video amplifier stage faulty
Data is displayed but focus is poor	Focus control adjustment
Focus anode supply defective
CRT defective
Data is displayed but brightness is low. Display size may increase and focus worsen as brightness is increased	Poor EHT regulation, horizontal output stage defective
Data is displayed but contrast is poor	Video amplifier or video output stage faulty
Contrast control out of adjustment
Data is displayed but brightness is low. Display size remains constant and focus correct as brightness control is varied	Incorrect bias voltage on CRT
No horizontal sync, display contains a number of near horizontal bars	Horizontal sync stage defective
No horizontal or vertical sync, display consists of rolling horizontal bars	Sync separator defective
Reduced height, good vertical linearity	Height control out of adjustment
Poor vertical linearity, height normal	Vertical linearity out of adjustment
Vertical foldover, abnormal height accompanied by poor linearity	Vertical oscillator, driver or output stage defective

Monitor stock faults 193

Action

the windings of the flyback transformer for continuity

If, in the above procedure, the signal at the collector of the driver stage is abnormal, check the driver stage and work backwards to the oscillator

Check CRT heaters for continuity. Check d.c. voltages at the CRT electrodes. Note that a high voltage probe will be needed to measure the final anode supply. If the d.c. voltages are abnormal, and particularly if any two of the voltages are identical, remove the CRT connector and check for shorts or leakage between the electrodes. Remove and replace the CRT if found to be defective. Take great care when handling the CRT since there is a risk of implosion if subjected to a mechanical shock

Check d.c. supply voltage to video amplifier stage. Check input connector. Check video waveform at input and work towards video output stage

Adjust focus control

Check d.c. supply to focus anode

Check for internal short between first anode and focus anode

Check d.c. supply voltage to horizontal output stage. Check d.c. voltages at CRT electrodes using a high voltage probe for the final anode measurement

Check horizontal flyback transformer for short-circuited turns.

Check EHT rectifier arrangement

Check d.c. voltages on video amplifier stages. Check video waveform at input connector and work forwards to the video output stage

Adjust contrast control

Adjust preset brightness control. Check d.c. supply to video output stage

Check waveforms at sync input and work towards the horizontal oscillator

Check waveform at sync input and work towards horizontal and vertical oscillators. Check d.c. voltages on sync separator stage

Adjust height control

Adjust vertical linearity

Adjust vertical linearity. Check d.c. supply to vertical stages. Check waveforms on vertical stages

	Vertical yoke defective
No vertical scan, display consists of a bright horizontal line	Vertical oscillator, driver or output stage defective Vertical yoke defective
Reduced width, good horizontal linearity	Width control out of adjustment
No horizontal scan, display consists of a bright vertical line	Horizontal yoke defective, linearity or width coil open circuit
Horizontal linearity poor, width normal	Linearity control out of adjustment
Horizontal linearity poor, width and brightness may be reduced	Horizontal flyback transformer defective, horizontal yoke defective

Typical adjustment procedure for a monochrome monitor

Adjustment	*Preset control*
Horizontal sync	H-hold preset resistor; H-osc. inductor
Horizontal linearity	H-lin. inductor
Horizontal width	Width inductor
Vertical sync	V-hold preset resistor
Vertical linearity and vertical height	V-lin. preset resistor; V-height preset resistor
Contrast	Contrast preset resistor
Brightness	Brightness preset resistor
Focus	Dynamic focus preset resistor; focus preset resistor

Check inductance of vertical yoke (typically 5 to 15 mH)
Check d.c. supply to vertical stages. Check waveforms and d.c voltages
Check vertical yoke for continuity (typically 2 to 10 ohm)
Adjust width control

Check horizontal yoke for continuity (typically 0.5 to 1.5 ohm). Check width and linearity coils for continuity (typically 0.5 to 2 ohm)
Adjust linearity control

Check horizontal flyback transformer and horizontal yoke for shorted turns (typical horizontal yoke inductance is 100 to 300 μH)

Procedure

Adjust H-hold preset to mid-position; adjust H-osc. inductor to centre of range over which picture achieves sync
Adjust for equal width character H at the left, right, and centre of the display
Adjust for correct width of display. (Note that this adjustment interacts with the linearity and it will be necessary to repeat the previous adjustment)
Adjust to the centre of the range over which picture achieves sync
Adjust V-height preset to obtain a display of approximately 70 per cent of the normal height
Use V-lin. preset to obtain equal height characters at the bottom, middle, and top of the display
Adjust V-height preset to obtain a full height display
Adjust for satisfactory display contrast
Adjust external brightness control to maximum (at which point the display raster should be clearly visible)
Adjust brightness preset to the point where the background raster just disappears
Adjust dynamic focus preset to minimum
Adjust focus preset to ensure that the display is focused uniformly
Adjust the dynamic focus preset for uniform focus at the edges of the display

196 Typical adjustment procedure for a monochrome monitor

Adjustment	Preset control
Screen centring	Centring magnet
Image deformation	Correction magnet

# Typical adjustment procedure for a monochrome monitor

Procedure

Repeat the two previous adjustments for optimum focusing

Adjust the magnet to provide a display which has the same periphery at the bottom and top, and at the left and right

Rotate the four correction magnets to gradually correct any deformation of the display

Index

Abbreviations:
  device coding, 29
  general, 9
  pin connecting data, 17
Access time, 100
Accumulator, 65
Acknowledge, 160
Address:
  mark, 113
  range, 62
ALU, 65
APL, 140
Architecture:
  microcomputer, 61
  microprocessor, 65
  PIO, 95
Arithmetic:
  logic unit, 65
  processors, 89
ASCII:
  character set, 148
  characters, 122
  control characters, 127
  to decimal conversion, 122
  to hexadecimal conversion, 122
Aspect ratio, 147
Assembly language, 139
Associative law, 56
Asynchronous:
  communications interface adaptor, 96
  transmission, 165
ATE, 162
Automatic testing, 162

Backplane, 152
BASIC, 140
Battery back-up, 103
Baud:
  generator, 171
  rate, 165
BDOS, 64
Binary:
  to ASCII conversion, 122
  to decimal conversion, 122
  to hexadecimal conversion, 122
BIOS, 64
Bipolar devices, 23
Bit rate, 165
Bit mapped graphics, 64
Boolean:
  algebra, 56
  identities, 56
  operators, 56

Boot ROM, 64
Buffer:
  memory, 100
  printer, 160
Bus:
  disk, 118
  GPIB, 162
  IBM PC expansion, 158
  IEC-625, 163
  IEEE-488, 162
  instrument, 162
  Intel Multibus, 156
  S-100, 152
  systems, 61
  transceivers, 61
Busy, 160

C, 141
Cache memory, 100
Capacitor colour code, 181
Cassette drive faults, 182
CCITT V24, 166, 169
CCP, 64
Cell configuration, 99
Centronics, 160
Character:
  font, 64
  set, 148
Charge coupled devices, 23
Clock:
  generators, 89
  real time, 90
CMOS, 23
COBOL, 142
Coding, 27
Colour code:
  capacitor, 181
  resistor, 179, 180
Colour mixing, 151
Commands:
  CP/M, 132
  GPIB, 162
  MS-DOS, 133
Commutative law, 57
Compiler, 137
Condition code register, 66
Configuration:
  memory cell, 99
  6502, 69
  68000, 85
  6809, 72
  8086, 82
  8088, 84
  Z80, 76
Control characters, 127

Index

Controllers, 89
  CRT, 89
  DMA, 89
  floppy disk, 116
  GPIB, 89, 162
  interrupt, 89
  keyboard, 89
  LAN, 90
  memory, 90
Counters, 90
CP/M, 132
  commands, 132
CPU, 61
  data, 88
CRC, 114
Crowbar, 60
CRT, 89, 144, 147
  controllers, 89
Current loop, 165
Cyclic redundancy check, 114

Data:
  address mark, 113
  communications equipment, 166
  rates, 169
  terminal equipment, 166
  transmission, 165
De Morgan's theorem, 57
Debugger, 138
Decimal:
  to ASCII conversion, 122
  to binary conversion, 122
  to hexadecimal conversion, 122
Decoder, 62
Decoupling, 59
Deleted address mark, 113
Demultiplexer, 62
Device coding, 27
Disk:
  cache, 64
  drive characteristics, 120
  drive faults, 184
  drives, 115
  8 inch, 108, 110
  5.25 inch, 108, 110, 120, 121
  floppy, 108
  format, 112, 114
  hard, 107
  media format, 110
  mini-floppy, 108
  3 inch, 109, 111
  3.5 inch, 109, 111
  Winchester, 107
Disk bus:
  Shugart, 118
  Winchester, 121
Disk operating system, 116
  CP/M, 132
  MS-DOS, 133
  PC-DOS, 133
Distributive law, 57
Divisors:
  of 255, 128
  of 256, 129
DMA controllers, 89
Double density recording, 112
Drive specification, 131
Drivers, 61
Drivespec, 131
Duplex, 166
Dynamic memory, 100

EIA standard, 166
Electrically alterable memory, 104
Erasable programmable memory, 103
Extension, 131

Fan-in, 54
Fan-out, 54
Faults:
  cassette drive, 182
  disk drive, 184
  monitor, 190
  printer, 186
Field, 113
File:
  name, 131
  specification, 131
Firmware, 131
Flag register, 66
Floppy disk controllers, 116
Flowchart symbols, 130
Format, 112, 114
FM recording, 112
FORTH, 143
FORTRAN, 143
Frame video, 150
Full-duplex, 166

Gate circuits, 27
GPIB controllers, 89
Graphic display processors, 89

Half-duplex, 166
Handshake, 160, 166
Hexadecimal:
  to ASCII conversion, 122
  to binary conversion, 122
  to decimal conversion, 122
High level languages, 138

IBM PC expansion bus, 158
ID address mark, 113
IN instruction, 92

Index

Indeterminate level, 56
Index:
  address mark, 113
  register, 66
Instruction pointer, 65
Instruction set:
  6502, 70
  68000, 87
  6809, 73
  8086, 83
  8088, 83
  Z80, 78
Instrument bus, 162
Integrated circuit technologies, 22
Intel Multibus, 156
Intelligent device, 62
Interface:
  audible transducer, 176
  CPU to ACIA, 96
  CPU to FDC, 117
  CPU to PIA, 93
  Centronics printer, 160
  LED driver, 176
  light sensor, 178
  mains controller, 176
  motor driver, 175
  optically isolated, 177
  parallel, 90
  relay drivers, 175
  S5/8, 172
  serial, 90
  serial current loop, 165
  stepper motor, 178
  switch input, 177
  temperature sensor, 178
  useful circuits, 175
Interfacing logic families, 60
Interlacing, 149
Interpreter, 136
Interrupt, 61
  controllers, 89

Karnaugh maps, 57
KERMIT, 173
Keyboard controllers, 89

LAN controllers, 90
Languages, 138
Line:
  driver, 170
  receiver, 171
Linker, 137
Listener, 162
Load, 55
Loader, 137
Logic:
  circuits, 24
  equivalents, 26
  gate characteristics, 54
  gates, 24
  levels, 56
  negative, 26
  positive, 26
LOGO, 143
Look-up table, 145
Low level languages, 138
LSI, 24

Magnetic:
  disk storage, 107
  recording techniques, 112
Mark, 165
Memory:
  access time, 100
  buffer, 100
  cache, 100
  capacity, 107
  cell, 100
  controllers, 90
  dynamic, 100
  plane, 144
  random access, 99
  refresh, 100
  static, 100
  wait state, 100
Memory map, 63
  CP/M, 64
  home computer, 64
  MS-DOS, 64
MFM recording, 112
Microcomputer systems, 62, 63
Microdrive, 109
Microprocessor architecture, 65
Mixed logic, 26
Modem standards, 172
Monitor:
  adjustment, 194
  faults, 190
MS-DOS commands, 133
MS-DOS, 133
MSI, 24
Multibus, 156
Multiplexed bus, 61

Negative logic, 26
Noise margin, 56
NMOS, 22
Null modem, 170
Numeric data processors, 89

Octal to hexadecimal conversion, 122
Operating systems, 130
OUT instruction, 92

Parallel I/O, 92
Parity bit, 169

Index

PASCAL, 144
PC-DOS, 133
PCB tracks, 60
Peripheral:
  control, 93
  data, 93
Peripheral interface adaptor, 92
Physical machine, 130
Pin connections:
  EPROM devices, 105
  FDC, 117, 118
  parallel I/O devices, 95
  RAM, 102
  regulators, 59
  serial I/O devices, 98
  Z80, 77
  4000 series, 49
  4500 series, 53
  6502, 69
  68000, 86
  6809, 73
  74 series, 37
  8086, 82
  8088, 84
Pixel, 144, 147
PL/M, 145
PMOS, 22
Pointers, 64
Positive logic, 26
Power supplies, 59, 108, 120
Powers of two, 122
Prefixes, 20
Primary channel, 166
Printer:
  faults, 186
  interface, 160
Program counter, 65
Programmable:
  memory, 103
  parallel interface device, 90, 92
  serial interface device, 90, 96
  sound generators, 90

RAM, 64, 99
  dynamic, 100
  static, 100
Random access memory, 99
Read only memory, 103
Read/write head, 116
Real time clocks, 90
Refresh, 100
Register, 65
  accumulator, 65
  condition code, 66
  control, 93
  data, 93
  data direction, 93
  flag, 67
  index, 66
  PISO, 97
  shift, 97, 145
  SIPO, 97
  stack pointer, 66, 67
  status, 67
Register model:
  6502, 69
  68000, 86
  6809, 73
  8086, 82
  8088, 84
  Z80, 77
Regulators, 59
Resistor colour codes, 179, 180
Resolution, 147
RGB video signals, 151
ROM, 103
  boot, 64
  electrically alterable, 104
  erasable, 103
  mask, 103
  programmable, 103
  system, 65
RS-232C, 165, 166, 169
RS-422, 167
RS-423, 167
RS-449, 168

S-100 bus, 152
S5/8, 172
Scale of integration, 23
Scanning, 149
SCART connector, 152
Secondary channel, 166
Sector, 113
  format, 113
Semiconductor memory, 99, 103
Serial data:
  format, 169
  support devices, 170
  transmission, 165
Shift register, 97, 145
Shugart bus, 118
Single density recording, 112
SLSI, 24
Sound generator, 90
Space, 165
Spikes, 60
SSI, 23
Stack pointer, 66, 67
Standard TTL load, 55
Start bit, 169
Status register, 66
Stepper motor, 179
Storage capacity
  comparison, 107
  floppy disk, 121

Index

magnetic disk, 107
  semiconductor RAM, 99
  semiconductor ROM, 103
  Winchester disk, 121
Stringy floppy, 109
Structured programming, 139
Supply distribution, 60
Support devices, 89
Symbols:
  ANSI, 24
  flowchart, 130
  MIL, 24
Synchronizing, 165
Synchronous:
  serial data adaptor, 96
  transmission, 165
System clock, 61

Talker, 162
Teletype, 165
Timers, 90
Track, 113
Track format, 113
Transceivers, 61
Transient program area, 64
Transients, 60
Transmission, 165
Tri-state, 62
TTL:
  input and output current, 55
  interconnection, 55
  load, 55

Unit load, 55
Universal:
  asynchronous receiver/
    transmitter, 96
  synchronous receiver/
    transmitter, 96
User RAM, 64, 65

Vectors, 64
Versatile interface adaptor, 92
Video:
  display processing, 144, 146
  display processors, 89
  memory, 146
  resolution, 147
  RGB signals, 151
  standards, 149
  waveforms, 150
Virtual machine, 131
VLSI, 24

Wafadrive, 109
Wait state, 100
Winchester disk, 107
  bus, 121
Write-protect, 116

Z80, 76
  configuration, 76
  family, 76
  instruction set, 78
  pinout, 77
  registers, 76
  support devices, 76

4000 series, 48
4500 series, 51

6502, 67
  configuration, 69
  family, 68
  instruction set, 70
  pinout, 69
  registers, 69
  support devices, 68
68000, 85
  configuration, 85
  family, 85
  instruction set, 87
  pinout, 86
  registers, 86
  support devices, 85
6809, 72
  configuration, 72
  family, 72
  instruction set, 73
  pinout, 73
  registers, 73
  support devices, 72

74 series, 29

8080 architecture, 67
8086, 81
  architecture, 67
  configuration, 82
  family, 81
  instruction set, 83
  pinout, 82
  registers, 82
  support devices, 81
8088, 84
  configuration, 84
  pinout, 84
  register models, 84

Also available from Heinemann Newnes

Servicing Personal Computers
£17.95
M Tooley

"A real gem. Some necessary theory is put over brilliantly . . . excellent chapters on equipment and trouble-shooting, and the packed book ends up with a magic reference section." *Computers in Schools*

"This book is worth having to test equipment usage techniques alone — the other information is a bonus." *Electronic Technology*

"Welcome indeed is Michael Tooley's exhaustive but easy-to-use book . . . it will go a long way to overcoming the awful problem of equipment faults." *Laboratory Equipment Digest*

"At last a crucial gap in the huge pile of microcomputer literature is filled — and this book fills the gaps so well it'll be hard to beat." *Education Equipment*

"A very worthwhile publication." *Television*

Softcover. 272 pages. Illustrated. 1985
234 × 165 mm. 0 408 01502 0

Newnes Radio and Electronics Engineer's Pocket Book

Seventeenth Edition

£6.95

Keith Brindley

Newnes Radio and Electronics Engineer's Pocket Book is an invaluable compendium of facts, figures, and formulae and is indispensable to the designer, student, service engineer and all those interested in radio and electronics.

"This little book must be a winner." *Hi-Fi News*

"Everyone involved in electronic engineering should have a copy." *Electronic Technology*

"Will certainly prove immensely useful . . . it cannot be rivalled at the price." *Radio and Electronics World*

'I only wish I'd known about this little book when I was struggling through college. Now, this review copy will never leave my side." *Electronics Times*

"Would have to be kept under lock and key to be retained." *What's New in Electronics*

"This splendid and unsurpassed pocket reference book cannot be too strongly recommended." *Elektor Electronics*

Hardcover. 176 pages. Illustrated. 1986
190 × 90 mm. 0 408 00720 6